M. A. Ronald.
10 Long Rigg
Clachan..

D0335326

THE HILL SHEPHERD

THE HILL SHEPHERD

Edward Hart

David & Charles
Newton Abbot London North Pomfret (Vt) Vancouver

ISBN 0 7153 7483 4

Set in 11 on 13pt Garamond
by Ronset Limited Darwen Lancs
and printed in Great Britain
by Biddles Limited Guildford
by David & Charles (Publishers) Limited
Brunel House Newton Abbot Devon

Published in the United States of America
by David & Charles Inc
North Pomfret Vermont 05053 USA

Published in Canada
by Douglas David & Charles Limited
1875 Welch Street North Vancouver BC

CONTENTS

To the hill farmers and shepherds throughout Britain whom I have visited, each and every one of whom shared their knowledge and hospitality without stint.

A good strong stick,
An intelligent dog,
A knife and piece of string,
An independence of thought
And a long free stride
To carry me across the ling.
To understand nature in all its degrees
From the heat of summer to the long deep freeze,
To come to terms with these varying moods
To enjoy the best and fight the cruel,
All hillmen know and accept these things
And take them in their stride,
To handle good stock and manage them right
Oh what pride!

Ray Dent, Shepherd of the Year 1976

ᴅᴍʙᴇᴍ·ALDERSON 77
THREE SCORE YEARS & TEN.

AUTHOR'S PREFACE

'When a man kicks a ball into a net or puts it down a little hole, there's a great to-do. If someone makes a good job of ploughing, works late every night to bring in the harvest or lambs a large ewe flock on his own, never a word is said.'

The words of an elderly Dales farmer passed through my mind whilst I was shaving one morning after a sheep breeders' meeting. There I had sat among men who braved all weathers day after day, sometimes out of sight of humanity, their sole concern the sheep under their care. Some owned their flocks and others were employees, but all had a characteristic dedication.

The dividing line between owner and employee is so indistinct as to be unrecognisable when the flock's welfare is in question. I recall one day at a Highland farmstead beneath a world-famous deer forest. The farmer and half a dozen shepherds with their dogs had congregated to discuss the day's work of gathering a particular hill. A stranger would have been hard put to decide which man of the group was the farmer, and which his shepherds. There was no lack of authority, but every man present was completely confident in his own and his dogs' abilities to gather that hill in his particular fashion.

News bulletins are seldom without a story of men on strike for the most trifling reasons; in one instance their car park was muddy. Here, at the other end of the scale, was a body of men and women embodying the old-fashioned virtues of dedication and loyalty. Awards for sportsmen, industrialists, beauty queens, actors and dress designers multiply year by year. Why not, then, a Shepherd of the Year?

As Northern Correspondent of *Livestock Farming* magazine, I knew something of the power of the farming press for spreading ideas. Yet more was required than an appeal to the agricultural community alone. Something was needed to catch the public imagination, to make the townsmen realise how much effort was involved in a way of life which he encountered only at holiday times. Luckily Kirkley Hall Symposium, a sheep farmers' get-together, was held soon afterwards at Northumberland College of Agriculture, Ponteland. At that 1972 meeting, I put some rather vague notions concerning a shepherds' award to Donald Dougal and Peter Weston of Pfizer Limited, and Marcus Oliver of *Livestock Farming*.

Yes, it was a great idea, this Shepherd of the Year, we all agreed, but how to get it off the ground? Who would be eligible, who would judge it, how would entries be processed and, above all, who would pay for it? Gradually a plan emerged to enclose entry cards in the magazine, and to select judges from different parts of the UK. Pfizer and *Livestock Farming* became joint sponsors.

Professor Gordon Dickson, Dean of Agriculture at the University of Newcastle upon Tyne, was asked to be chairman, backed by Northumberland vet Gerald Curry. From north of the Border, 'Billy' Elliot could be relied on to combine his chairmanship of the British Wool Marketing Board with furthering the scheme, while Anglesey sheep farmer Iolo Owen was well placed for Wales.

Plate 1 Shepherd of the Year Award statuette

11

One of the objects of the award was to improve the shepherd's image with the general public. Another was to encourage young people into the profession. A shortage of shepherds has undoubtedly led some farmers to give up or reduce their flocks, and has deterred others from keeping sheep. Shepherd of the Year was launched early in 1973, and has proved a continuing success story. All those involved in its organisation feel that the award has fulfilled those initial aims. Farmers and others responded by entering their shepherds, or those of their neighbours. The winner has been filmed on his or her own ground each year, and so part of the shepherd's work has already reached a wide audience. This book aims to fill in the details.

WINTER ON THE HILLS

1 AN INTRODUCTION TO THE SHEPHERD'S YEAR

It is winter on the hills. The white bowl of a great valley glistens under a blue sky edged with inky, threatening clouds. Against the snow a dark ribbon is etched thinly, a line of sheep moving purposefully in single file through the deep snow. They follow-my-leader towards a solitary figure who, from easy and confident stance, directs two black-and-white dogs in rounding up the stragglers, while a third dog noses the sheltering wood near at hand.

As the sheep reach the home paddock at steady pace, the figure moves towards them and slings flaps of hay from bales he has carried from the nearby barn. He finishes his task with 200 hill ewes milling around, calls off the dogs, and simply watches. This man is a hill shepherd; these and another 1,100 sheep are his charges; his only help his son. Hatted, wearing mittens, thick trousers and leather jerkin against the keen air of the 1,500ft

13

contour, he resembles several hundred such professionals who daily tackle highly responsible jobs.

The sheep are likely to be dark-faced and horned, or light-faced and hornless. They have been bred for generations to winter on Britain's hardest hills, fending for themselves most of the time. When the deep snow is frozen, even their scraping hoofs cannot penetrate to the scant vegetation below, and they are fed hay into which is tucked the goodness of a June day. The hay rustles as they tear at it, disappearing into hungry mouths at a surprising rate. The sheep move from one pile to another, always seeking a tastier piece than they are consuming, always keeping a wary eye on the dogs to whose authority they reluctantly submit.

Still the shepherd continues to watch his flock. It seems an easy, pleasant life, out there on the silent hills from which snow storms have barred even a passing car—no tube trains, no rush hour, no boss within five miles. Yet into that penetrating gaze goes years of accumulated experience, noting immediately signs of lameness, blindness, weakness, any departure from the norm.

Two days previously, the shepherd and his son had worked until dark, prodding long sticks into snow drifts the height of the stone walls they faced. One dog, Nell, helped them. She is a Border collie with a peculiar aptitude for 'setting' or smelling out buried sheep, much as a well bred gun dog 'points' at its quarry. On two occasions, Nell dug ferociously, and the men's efforts with shovels at that spot resulted in the release of two sheep from a snowy grave. Yesterday, three more had returned to the flock, having themselves scrambled free from the drifts.

Far below, a stone house stands grey against the surrounding white. A wisp of smoke curls from its chimney, and inside a meal is being cooked for hungry men. A still day such as this is seldom a bad one on the hills; gales and lashing rain form the worst weather combination in an area where shelter of any sort is scarce.

That morning the shepherds had risen at 7 am, fed the cattle in the yard and given hay to the hoggs—next year's replacement ewes. Then came breakfast of bacon and eggs, with several

Plate 2 Winter in Weardale. Ray Dent's son Ian slinging flaps of hay to a section of the Swaledale flock. This homestead was once a farm on its own, then housed a shepherd and, like many in the Dales, is now used for stock only, as the smaller farms have been amalgamated. 'Snow blindness' is a condition for which the shepherd must be constantly alert, the eyes of infected sheep being dressed with a veterinary ointment. If untreated, the animal cannot see to eat properly, becomes emaciated and dies

telephone calls interrupting the meal, for a shepherd's house on the 1,250ft contour is nevertheless a business HQ. Two strayed sheep had been taken up by nearby farmers. 'Good, straight-forward neighbours', said the shepherd. 'Life would be grand if everyone were as helpful. We can always rely on getting our sheep back from there.'

From then on the various lots of ewes are fed their daily hay ration of 1lb or more, depending on weather and snow cover.

Batches of more than 200 are unsuitable, as the stronger rob the weaker before they have time to feed. In the short winter after-noons any stray sheep are fetched home. Soon it is time for evening feed for the hoggs and cattle, then tea in the welcoming kitchen and, with luck, an evening's relaxation ahead.

If the winter landscape is green rather than white, the shep-herd's task is much easier. Hay may still be fed, but tractors can travel right to the various lots, and consumption is much reduced. In any spare time, gaps in the dry-stone walls are mended—a never-ending task—a length of road repaired, buildings tidied and manure led onto the fields. For hill farming is an occupation in which periods of strenuous activity alternate with slacker periods used to catch up on jobs necessarily neglected during the busy times.

Late in April comes lambing. Timed to coincide with an-ticipated grass growth, the date for each particular farm and county has been arrived at by trial and very expensive error over the decades. An early spring is a rarity above the 1,000ft level, where winds in April and May can be bitingly sharp. 'It's a good thing the lambs are in a warmer spot than this', say the shepherds as April hailstones bounce against the woolly backs of ewes heavy with late pregnancy.

For a month at lambing time, work from 6 am to after darkness has fallen is the invariable rule. Then there is no respite and 'lambing snows' are but one hazard. Yet eventually this hectic season of births gives way to easier, early summer days when work among a healthy flock is pleasure. The next busy time is after Midsummer Day, when shearing and haymaking inevitably clash. Even where all winter fodder is bought and none made, the longest days of the year are filled with activity as one lot of ewes and lambs after another is driven painstakingly down to the shearing sheds, then slowly back again to the green, wall-encircled hills.

By August the lambs are almost as big as their dams. Now they are weaned, some are sold and the rest kept as flock replacements next year. The older ewes also go, and the whole autumn period

Plate 3 Summer in the hills. Lush vegetation, miles of dry-stone walls, sheep moving steadily in front of the dogs, and shirt sleeves constitute an idyllic picture of sheep farming. Sheep here are Dalesbreds and Swaledales, returning to pasture after shearing, lambs at foot. The walled enclosures stretching up the hillsides are termed allotments, being allotted to holders of lower fields with grazing rights on the open moor above

is a succession of sales, journeys and bustling company.

Then comes 'tup time', in many respects the real start of the shepherd's year. That strange hormone action that links oestrus in the ewe and libido in the ram with the shortening days comes into play. Autumn is the natural time for most breeds of British sheep to mate, and to the shepherd the season is as vital as lambing time. It is his job now to see that every ewe has the chance of a mate. If she is not mated, she cannot carry a lamb through the winter storms and lamb on the hillside when April comes again. She would make no profit, and hill shepherding, like any other business or profession, is carried out primarily for profit. Early in the New Year the tups are 'taken up', and the ewes are joined into manageable lots for the winter where they can best be fed and observed by the shepherd on his lonely hillside.

HERDWICK.

D·H· & E·M·ALDERSON
1977

2 THE FLOCKS HE HERDS

Though it is often said that dark-faced, horned sheep have grazed the hills of northern Britain since time immemorial, the simple truth is that we don't really know for how long. There are a number of such hill breeds with rather similar general characteristics, each of which has evolved to suit the conditions of its own particular hill country.

Most numerous is the Scottish Blackface. Comprising 30 per cent of Britain's purebred sheep (there are a great many halfbreds and crossbreds) the 'Blackie' is found at all elevations, from sea level to the summits of the highest Scottish 'bens', though most Blackface flocks graze the hills to some 2,000ft. She may live on grazing so sparse that several acres are needed to support her, but the 'Blackie' usually finishes her days on kinder conditions lower down the hill. There, crossed with the Border or Blue-faced Leicester, she produces the Greyface, an excellent dam for lowland fat lamb production.

Wool from the Blackface is in a class of its own and in great demand for the Italian mattress trade. Overcoats, heavy blankets and working tweeds are made from the finer wools, for this breed boasts several distinct types of fleece according to its surroundings—Perth, Newton Stewart and Lanark types are each recognised by the breeder. Much of the best Blackface wool goes to the Islands to be woven into the famous and enduring Harris tweeds.

Though Blackfaces are found in Northumberland, Cumbria and the North York moors, the main hill breed of northern England is the Swaledale. Horned like the 'Blackie', instead of the broken black-and-white face markings this sheep has a snow white muzzle and jet black face. At least that is the ideal, but old Swaledales tend, like humans, to go grey with age, though they remain very active sheep.

When the late John Teasdale farmed at Helmsley, North Yorkshire, he bought some Swaledales from the nearby hills. As the sheep were adept fence breakers, John's sons fenced the home paddock so securely that 'an old hare couldn't get out', but this did not prevent the wandering ladies from tramping round and round the perimeter for three days, wearing a track in the process. They then discovered, or created, a weak point, and disappeared to the heights whence they came. Yorkshire farmer Maurice Metcalfe, who introduced Swaledales following a disease outbreak, found that they climbed to the top of Buckden Pike's 2,000ft of their own accord. It is the combined foraging and milking ability of the Swaledale ewe that has helped her spread into Scotland, Devon and even Brittany.

From the western slopes of the Lakeland mountains to the moors bordering the North Sea, you will find Swaledales. The Pennine dales are thick with them, and the show at Tan Hill— England's highest pub—in late May is not to be missed. The breed takes its name from the valley of the Swale, a Yorkshire river which rises high on the summits of remote fells near the Cumbrian boundary.

The Association's description reads:

The indigenous sheep which had been able to eke an existence from these starved and acid fells—where heather and draw moss predominate—were the raw materials in which the hill farmer of the past took great interest and, carrying on where nature left off, they sought to improve the type of sheep and yet maintain those strong qualities which had been developed in the animal under its natural environment. Carefully they selected and bred, until there emerged the remarkable Swaledale that we know today.

Self-reliance is an essential quality in a hill-sheep breed, and none needs it more than the Swaledale. Several times I have driven down from Argyllshire and south-west Scotland, through Brampton and England's highest market town, Alston, to find snow and fog only on the Durham hills around Wearhead and Nenthead.

The Swaledale Flock Book was founded in 1921, and there are now almost 1,000 members, mostly active breeders. The breed organisation is divided into five districts, lettered A–E, each with its own secretary and assistant working under one head secretary and staging its own annual ram sale. Castleton, Hawes, Kirkby Stephen, Middleton-in-Teesdale and St John's Chapel are the great sale centres. At Kirkby Stephen alone, over a thousand Swaledale rams change hands. After three days there, the visitor may wonder if there is any other human activity to compare with breeding Swaledale rams!

Wool from these sheep was formerly used for carpet-making and blending, but the Bradford Institute of Wool Technology has carried out further experiments which have resulted in hard-wearing yet lightweight fabrics suitable for both men's and women's wear. There is also a knitting wool produced solely from Swaledale fleeces. National Sheep Association and British Wool Marketing Board stands at the big shows are ideal places at which to see the various wools and cloths.

One great value of the Swaledale, however, is as dam of the Mule, with its face of speckled brown or grey and its large numbers of fast-growing lambs. Sire of the Mule is the Blue-faced Leicester, small flocks of which are found on many northern upland farms, kept solely to provide males to mate with the older Swaledale ewes.

A breed found mainly in Lakeland is the Herdwick, sharing with others a legendary association with the wrecked Spanish Armada, but a more probable one with Scandinavian sheep. Herdwick mutton was certainly eaten at the present Queen Elizabeth's coronation feast. Lambs of the breed are jet black, turning grey with age, the colour of John Peel's hunting jacket. Because of this range of shades, multi-coloured wool is obtained without dyeing.

Polled in the female, usually horned in the male, the Herdwick has one characteristic unique among British sheep. It lies low when the dogs are rounding up the flock, and wily old Herdwick ewes have been known to miss several 'gathers' simply by hiding behind rocks and refusing to run with the rest. For this reason a barking dog is an asset, and perhaps the newly introduced New Zealand huntaways, which bark on command, may find a niche among the towering Lakeland fells.

The breed takes its own line in two other ways. Herdwick rams are hired as well as sold at the Keswick ram hirings each September (see Plate 5), and the sheep are counted in Cymric numerals, which roll off the tongue like a well bred Border collie loping up a hillside. Yan, tyan, tethera, methera, pimp, goes the rhythmic roll-call. Sethera, lethera, hovera, dovera, dick, it continues. Eleven is yan-a-dick, twelve is tyan-a-dick, then tether-a-dick, mether-a-dick, bumfit. Yan-a-bumfit, tyan-a-bumfit, tether-a-bumfit, mether-a-bumfit, giggot, takes the counter to twenty. When the shepherd has reached twenty, he puts up one finger and starts again. When he has five fingers up he has reached a hundred and puts a pebble in his pocket. Science brings us computers and calculators, but the Lakeland shepherd excels in beautiful simplicity of method.

Plate 4 First winner of the Shepherd of the Year Award, Mrs Dorothy Wearmouth (formerly Miss Dorothy Bell), with one of her Swaledale gimmer hoggs

Other North Country hill breeds are the Dalesbred, Derbyshire Gritstone, Lonk, Rough Fell and White-faced Woodland. An offshoot of the Swaledale, the Dalesbred has a fleece with more crimp or pearl, and is ideally suited for mating with the Teeswater or Wensleydale to produce the ever popular Masham.

Derbyshire Gritstones are big, powerful sheep of the south Pennines, where industrial pollution is added to natural hazards.

On their native heath, amidst ling and furze, rocks and boulders, they are exceptionally hardy and wiry, possessing a priceless immunity from ills that palefaced sheep are heir to in the lowlands, and endowed with the persistent energy common to denizens of bleak and ruthless mountainous districts

Plate 5 Herdwicks at the Keswick September ram hirings. Some animals are sold, others hired for the season and returned to their owners at a similar event next May, where prizes are given for the best kept rams. As with most hill-farming events, the hirings combine business and pleasure, for hill people relish the chance to exchange news and views with neighbours whom they seldom see. The Herdwick's wool is John Peel grey, and very weatherproof

Fellside (1606). Lambed 1919.
Owner and Breeder : John Dalton, Fellside, Caldbeck.

Fellside (1607). Lambed 1919.
Owner and Breeder : John Dalton, Fellside, Caldbeck.

Fellside (1608). Lambed 1919.
Owner and Breeder : John Dalton, Fellside, Caldbeck.

Fellside (1609). Lambed 1919.
Owner and Breeder : John Dalton, Fellside, Caldbeck.

Hayknot (1610). Lambed 1919.
Owner : C. J. Pears, Fellside, Caldbeck.
Breeder : W. Pears.

Scar Sparkle (1611). Lambed 1919.
Sire Derwent Sparkle (115). Dam Flock No. 70.
Owner and Breeder : Robert Teasdale, Branthwaite, Caldbeck.

Low Pike Ruler (1612). Lambed 1915.
Owner : Daniel Tyson, Low Fellside, Caldbeck..
Breeder : J. Hawell.

Low Pike (1613). Lambed 1919.
Owner : Daniel Tyson, Low Fellside, Caldbeck.

Low Pike Mountain (1614). Lambed 1919.
Owner and Breeder : Daniel Tyson, Low Fellside, Caldbeck.
Breeder : A. Parker.

Hudscales High Pike (1615). Lambed 1919.
Owner and Breeder : John Teasdale, Hudscales, Hesket-New-Market.

Hudscales Fuzzy (1616). Lambed 1919.
Owner and Breeder : John Teasdale, Hudscales, Hesket-New-Market.

Hudscales Peter (1617). Lambed 1919.
Owner and Breeder : John Teasdale, Hudscales,, Hesket-New-Market.

Hudscales Sambo (1618). Lambed 1919.
Owner and Breeder : John Teasdale, Hudscales, Hesket-New-Market.

Sample entries from the Herdwick Flock Book

This is a true description of the Gritstone, a polled sheep, used in experiments to breed other hornless sheep in an attempt to combat sheep head fly (see Appendix A), against which no effective solution has been found since dieldrin was banned as a dip. Their short, dense fleeces dry out more rapidly than the trailing coat of a Blackface. The Lonk has similar fleece and black-and-white face and legs, but has sweeping horns in the ram, strong and flat horns in the ewe.

Further northwards around Kendal, Cumbria, is the home of the Rough Fell, a grand, quiet breed of hill sheep and a favourite with shepherds. Its wool is straight and long, and its black-and-white face markings resemble the Scottish Blackface, both sheep probably being descended from the same Blackfaced Heath breed, of which Culley wrote in 1794: 'This hardy wild-looking tribe are first met with in the North-West of Yorkshire, and are in possession of all that hilly or rather mountainous tract of country adjoining the Irish Sea, from Lancashire to Fort William.'

The White-faced Woodland is said to have been traced back to the Derbyshire Peak District since medieval times. It is now regarded as synonymous with the Penistone sheep, two of which once escaped from Kentish pastures after being sold in that area, and walked back home. To commemorate this feat their horns were hung in the local church in the village of Hope, Derbyshire.

The way which the little Welsh Mountain sheep are crowded onto steep, improved pastures is an object lesson for the rest of the world. Second in numbers only to the Scottish Blackface, the Welsh Mountain has a similar divergence of type, and almost thirty different kinds have been classified.

Horned in the male but not in the female, this ancient breed is small in size but capable of producing a lamb growing almost to the size of its dam before weaning at about four months of age. These big lambs are sired by another breed, often the Border Leicester, and are known as Welsh Halfbreds. (One of the success stories of post-war sheep production.) Each year about 450,000 Welsh Mountain draft ewes leave the hills of Wales for the lowlands, and lambs produced under this system frequently

Plate 6 There are many different types of Welsh Mountain ewe, some having more tan on face and legs than these being driven from a mountain road. Note the fleece-marks on the lambs, stamped by the shepherds with branding irons dipped in coloured fluid. These lambs are old enough to return to the mountains; they are generally born in more accessible and relatively sheltered enclosures lower down. At this age the lambs have not learnt to respect the dogs, and charge hither and thither causing a great deal more work for the hill shepherds and their watchful Border collies

figure at London's famous Smithfield Show each December. Stringent rules keep quality high at the official collective sales of Welsh Halfbreds, and the Field Officer travels many thousands of miles each year to check that new customers are happy with their choice.

The Welsh Mountain Sheep Society was formed in 1905 and today has almost 700 members. In almost every flock of Welsh Mountain sheep some will be found to have rusty-yellow, or tan, faces and legs, while the remainder will possess white faces and legs. As pointed out by John Roberts in *Livestock of the Farm*, in some districts the tan faces are predominant, while in other districts the reverse is true. And the *Standard Cyclopedia of British Agriculture* credits this 'distinct race' with having 'played the most important part in the economy of the country for many centuries'. Today performance and progeny testing of rams is carried out in association with the Welsh colleges, whose staff work very closely with the farmers.

The Black Welsh Mountain is a related breed, selected as a separate strain for perhaps a century, and having its own Flock Book since 1922. In the Middle Ages, the black wool of the Welsh Mountain sheep, known as 'Cochddu', commanded a premium, and is still a valuable commodity today. In combination with white wool, it makes cloth of pleasing checks and patterns. This hardy breed has been exported to several countries.

The Welsh Halfbred was evolved as a competitor to the long-established Scottish Halfbred, the product of the same Border Leicester ram out of a Cheviot ewe. Cheviot sheep originated in the wild Border hills of that name, and are white of face, fleece and legs, always hornless in the female and often in the male. The breed has now split into two, the larger North-Country Cheviot being better adapted to the more fertile lower hills, while the South Country or Hill Cheviot thrives on rather sparser fare. No other sheep has suffered so much at the hands of forestry companies who, despite their protestations to the contrary, secured much of Scotland's best grazing lands, especially in the south.

Lockerbie and Hawick are the main Hill Cheviot centres. At the former town the Western Cheviots are sold, some rams being horned, whereas in the eastern Borders all are polled.

Our true hill shepherd is more likely to be concerned with the prick-eared 'Southie'. It has the best mutton conformation of any hill breed, though the Exmoor Horn ranks high. As Dr Allan Fraser in *Sheep Husbandry* writes: 'There is no more beautiful sheep than the Cheviot found on its native hills. It has a carriage, alertness and keenness of expression unsurpassed by any other breed.'

No animal or breed is perfect for all situations. The Cheviot's broad back makes an ideal butcher's sheep, but calls for extra vigilance from the shepherd. As the white fleece lengthens in spring, the sheep is liable to become cast on its back, 'couped' as they say in the Borders, or 'rig-welted' in north of England phraseology. A ewe heavy in lamb is especially liable to be trapped in some hollow in the ground in this way, and will die unless quickly spotted. New Zealand sheep dogs are taught to help sheep to their feet by tugging at the shoulder wool. Once the sheep is sheared, the danger recedes. Active sheep like the Swaledale are seldom trapped in such a fashion, but they cannot claim to be in the same class for meat.

The following extract is from Culley's *Observations on Livestock* (1794). He describes Cheviots as

. . . hornless, the faces and legs in general white; the *best kinds* have a fine open countenance, with lively prominent eyes; body long, fore-quarter wanting depth in the breast, and breadth both there and on the chine; fine clean small-boned legs; thin pelts; weight of carcase when fat, from 12lb. to 18lb. per quarter; fleeces from 2½lb. to 3½lb. each; and sold in 1792 for 11d. per lb.

The wool is not all fine, there being in a fleece of 3lb. weight, only 2lb. of fine wool, worth 1s. per lb. (when the whole fleece sells at 10d. per lb.) and 1lb. of coarse wool worth only 6d. per lb.

Some of the Cheviot sheep are speckled on the face and legs, but those are, probably, a mixed breed, from crossing at different times with the heath sheep, to whom they have been long neighbours; for, as you leave the heights of Annandale to the Eastward, you insensibly lose the Heath sheep and mixed breed, after which, all those extensive fine green hills on the Scotch and English Borders, (extending from Reedwater, on all sides the mountains of Cheviot, to the barren Heaths of Lammer-moor) are covered with the Cheviot Breed.

Exmoor and Dartmoor were improved by Scots colonists who naturally took with them their Blackface sheep. Strains of Blackface graze there still, but the local hill breed is the Exmoor Horn, an attractive sheep with superb mutton conformation, blocky and well fleshed. When winter gales roar in from the Atlantic, sheep have to be tough to survive on those moors so beloved by the summer tourist. The White-faced Dartmoor is another breed kept on the Moor from May to September, grazing to 2,000ft.

SCOTTISH BLACKFACE RAM

3 THE HILLS HE WALKS

One third of Britain is upland. Part of this vast acreage is improved grass, but England has one million acres of common land, and Wales half a million, most of which is in its natural state and unsuited to any purpose other than sheep grazing. There is, however, no common land in Scotland. Deer forests take a big chunk north of the Border, and sporting interests predominate on some estates to the detriment of more useful stock, but sheep farming remains a prime basic industry. Three-quarters of Scotland's agricultural land is rough grazing.

It is foolish to generalise about lowland farming, but criminal to generalise about hill farming, says Lakeland sheep breeder George Wilson who, with his brother Norman keeps Herdwicks and Swaledales. Their farm, Glencoyne, is owned by the National Trust, but has been farmed by the Wilson family for many years. Its lower land borders picturesque Ullswater, with a constant flow of summer traffic making stock movements dangerous. From Glenridding village Glencoyne's boundary turns inland to the watershed, where Raise, 2,889ft, Sticks Pass and The Dods, touching 2,807ft, entail a two-hour walk from the sheltered homestead before gathering can even begin. The northern boundary comes down Hartside and Birkett Fell, the latter named after the late Lord Birkett who defended Ullswater against Manchester Corporation's reservoir schemes, and who died tragically the weekend after victory. Previously the fell had been called Nameless.

One lambing paddock at Gowbarrow is used principally for the Herdwick ewes in-lamb to North Country Cheviot tups. Pure-bred Herdwicks and Swaledales make up the rest of the 2,600 ewe stock. During one recent lambing, Glencoyne's staff of three was reduced to one overnight when Norman Wilson succumbed to appendicitis and shepherd John Maxwell damaged his leg in a tussle with a Swaledale tup. Such is the comradeship of hill farming that the neighbours rallied round, although they themselves were up to their necks in work.

Swaledales are also the breed kept by the 1976 Shepherd of the Year, Ray Dent, who works at Glenwhelt, Weardale in Co Durham. Neither Glenwhelt nor the other hill farms described here may be called typical, for there is no such thing. But they do demonstrate something of the diversity of hill farming conditions, and the differing situations with which a hill shepherd might be faced.

The homestead, right by the roadside on the 1,250ft contour, is by no means as isolated as many hill farms. It is less than two miles from the village of Daddry Shield, and looks down onto Weardale with its multitude of houses and farms. The buildings straddle the road, older sheds and sheep yards on one side, new erections and stores on the other. The farm serves as a Mountain Rescue Station. Calls are infrequent, as Glenwhelt is some miles from the Pennine Way where most troubles occur, but one SOS did come while Ray Dent was in London receiving his award!

Glenwhelt has 600 acres of enclosed grazings and almost 3,000 of rough grazings, some of this being unfenced. Lack of fencing makes a great difference to shepherding; hill sheep are trained or

'hefted' (see Glossary) onto a certain stretch of territory, each ewe lamb learning the boundaries from her mother, but a certain number fail to stick to the rules.

The highest point on Glenwhelt tops 2,280ft, with much of the land above 1,500ft and the snow line. Springs are late, summers short, winters often severe, and wind and hill fog are constant weapons in nature's armoury against man's settlement of the heights. The farm forms a huge basin in the hills, levelling out on the tops to form the bleak plateaux of Chapel Fell and Swinhope. Swinhope Head forms one heft, Swinhope Fell two, being split by Swinhope Burn into north and south sides. Chapel Fell starts at the 1,800ft level, standing to the south-west of the steading and with tremendous views of the surrounding dales and across towards the Cumbrian Mountains. Another heft of Swaledales runs on West End, and then the fell gives way to the intakes, land literally taken in from the moor.

Most of these intakes are on the east side of the road winding up the valley. Fell Bottom's 70 acres adjoin the slightly smaller Dam Pasture, so named because the dam from a disused mine stands there. One pasture, called simply Enclosure, is 60 acres, and stands next to Camp Lot's 160 acres, where Methodist camp meetings took place around the time of John Wesley's tours—Methodism is still a force in the dales. The Park's 60 acres have a number of useful trees and Long Field is of similar size. East Lot is a separate enclosure right on top of the 2,000ft line, with poor walls round it. Good Lot has 180 acres, while Greenley Hills' 240 acres near the homestead are used for hogg wintering. Better fields graze crossed or halfbred lambs, because of the extra growth potential and at Glenwhelt, Curragh's Lot, Straight Lot and John Emerson's 'allotment'—walled enclosure—are used for this purpose.

Ray Dent followed his father George as head shepherd of these acres. His son Ian is assistant shepherd, while his wife Lena plays a vital part, helping at lambing and other busy times as well as dealing with the telephone and providing that much-needed endless succession of meals. The owner of Glenwhelt is John Vickers of Tow Law, whose father employed Ray Dent's father. 'I love this place', he says. 'I have no need to come, though my home farm is seventeen miles away. When I have been incapacitated through illness, the work at Glenwhelt goes on just the same.'

In the Rob Roy country near Crieff, in the southern Grampians, lies one of the farms of Ben Challum Ltd. The road comes to an end in this haunt of ravens, black-hooded crows and red deer. Snow-capped mountains block the way out, yet the drovers of old crossed these heights with their coast-bound stock. Ben

Challum, 3,354ft, rises at the far end of the glen. Nearby is the Herring Stone, scene of bartering for cattle with fish, grain and other things.

The farm at Glen Lochay is divided into seven hirsels (see Glossary), and totals nearly 24,000 acres. Six hirsels carry 700 sheep and 15–20 cattle each, and a smaller one has 300 sheep. The stock is comprised of nearly 4,000 Blackface ewes, and the cattle are Luing or cross-Highland. Two hirsels at the top of the glen are ranched, while the others each have a shepherd in charge. There are three generations of sheep always in the glen, after which they are moved to kinder conditions.

At Cononish Farm, Tyndrum, Crianlarich, Perthshire, John Burton runs a ewe stock of 1,750, mainly Swaledale/Blackfaces. The farmstead is at the end of a track requiring twenty-five minutes of careful Land-Rover negotiation. Standing at a 950ft altitude, this isolated house is five miles from the farm's furthest boundary, the holding's 5,500 acres including three bens, or mountains over 3,000ft. A further 4,000 acres are rented.

Rainfall at Cononish is 100in a year. Whoever coined the phrase 'this small, overcrowded island' to describe Britain had presumably not left London, and had certainly never herded Cononish. Spring weather is so harsh that lambing is delayed till 26 April, and one hirsel formerly kept a stock of wethers only (see Glossary). One man is employed, and the sheep must largely fend for themselves, even at lambing time.

'We have had a very bad spell of stormy weather for tupping time', said John Burton. 'We did not manage much herding this time because of the floods, mist and gales. However, no doubt there will be some lambs, as we have seen it as bad in the past, and still survived.' The young lad employed at the farm has now left, and it will not be easy to get a replacement as shepherds are almost impossible to find in that remote area.

Today's sheep farmers are not the first to suffer labour problems. In *A Hampshire Shepherd*, published in 1902, H. Rider Haggard wrote:

[George Piper, a shepherd of over seventy years of age] stood in the cold wind, upon the bleak down crest, watching the fold much as a dog does, and now and again passing the hurdles to do some little service to his flock, every one of which he could distinguish from the other. This, too, on a Sunday, the day on which it is so difficult to keep the modern stockman to work, however necessary.

At three Stirlingshire hill farms of lower elevation, Townhead, Lurg and Craigton, in Fintry, James McEwan runs 2,000 Swaledale/Blackface ewes. The bulk of them are crossed with Blue-faced Leicester tups to breed the Greyface. On a total of 2,500 acres 400 head of cattle are summered.

Taking the tups to the ewes is a major operation on farms of this scale. Land-Rover and trailer now simplify the task, saving time, wear and tear compared with the succession of long treks previously necessary.

On these lower hills a lambing percentage of 130–150 is achieved, whereas on higher ground the shepherd is well satisfied with one lamb per ewe. More may be an embarrassment; ewes with twins cannot be summered on the hill's natural grazing, must be kept inbye (see Glossary) where they compete with cattle or eat the next year's hay crop. James McEwan uses solid block concentrate feed. Hard, circular blocks are simply set out on the hill at weekly intervals, and the sheep gnaw at them when they wish. This method is a help at lambing time, for ewes leave their young lambs in the rush for the daily ration of nuts and cubes, and mis-mothering is always a possibility. Ewes and lambs identify each other by sound and smell rather than by sight. 'It would be a clever man who could sort this lot out!' said one North Country hill farmer when a mass of ewes joined the scramble for appetising cubes, leaving their young lambs in uneasy, discordant bunches. Yet, as usual, the sheep soon coupled up and trotted off contentedly in appropriate pairs.

Cheviot sheep have been kept on The Lee, Longframlington, Morpeth, Northumberland, for a hundred years. The present occupier, Michael Aynsley, is grandson of the breed pioneer, and down the years the farm's basic policy has changed little. 'The original stock was typical Border Cheviot, as bred on the high ground there', explains Aynsley. 'We have selected a slightly bigger sheep for our own hills, to breed the size of Scottish Halfbred ewe the customer wants.' These Scottish Halfbreds are sired by Border Leicester rams out of the Cheviot ewes. Of The Lee's 830 Cheviot ewes, 300 are bred pure to provide flock replacements, the balance being mated to Border Leicesters.

Cheviot tups have always been bought from the best stocks, including Attonburn, Black Haugh, Cliftoncote, Elsdonburn and Milsington. A good big sheep with character is sought, one that looks alert and will go to the hill and do a good job. It needs a dense coat, and plenty of white hair on the forehead as a protection against head fly.

Within sight of the North Sea and overlooking the coastal plain, the hill ground has natural shelter in the form of quarries. Whinstone for roadmaking has been led away, and the quarries afford a free-draining floor for the hill cows fed there. A transport box with weldmesh sides is mounted on the tractor to reach any part of this particular hill farm, and carries nine ewes. Each is marked as she is loaded, and the lambs marked in similar sequence to prevent any mix-up at the other end.

Not far north of The Lee is Redesdale Experimental Husbandry Farm at Rochester, Otterburn. Standing on the eastern slopes of the north Pennines, it has over 4,000 acres of land, some of it improved hill. Most of the sheep are Scottish Blackfaces.

Communications are a first step to hill land improvement. On some types of hill it is fairly easy to bulldoze a track resting on stone or rock; in other situations, there is no such supporting substrata within range, and the soil swallows apparently endless loads of hardcore before stability is gained. Peat over clay provided a puzzle at Redesdale, the peaty topsoil necessitating a constant filling of ruts where roads were made by conventional methods. Now a new civil engineering fabric has proved successful. Easy to cut but hard to tear, long-lasting in situ, it is capable of being taken up and relaid provided exposure to light is avoided. In farming practice, it is covered with 4in of stone.

Across the Pennines at Great House EH Farm, Rossendale, Lancastrian Harry Mudd directs a smaller hill farm as befits the populated valleys edging the industrial zones. Here the local Derbyshire Gritstone is one breed of sheep used, and experiments are carried out to fit an early lambing flock into the hill farming system with Suffolk and Dorset Horn ewes.

In the South West, Liscombe EHF, Dulverton, Somerset, tries out upland sheep policies under Jack Willcock's directorship. Somerset and Devon are intensively stocked sheep counties, North Devon being reputedly among the most thickly populated sheep areas in the world. A favourable climate in an area of sound economy, backed by the mixed blessing of tourism, lends hope to the farmers of the South West, but there are no easy prospects for the men of Wales.

At the Ministry of Agriculture's Welsh station, Pwllpeiran EHF, Director Meredydd Roberts and his staff battle with 100 inches of rainfall and steep and difficult terrain. In afforested areas, carrion crows and other vermin add to the hill farmers' problems, and one year no sets of twins were reared at Pwllpeiran because of carrion crows.

One thing is certain, life on the hills is never easy. Travelling in Britain's uplands shows their still untapped potential, but frequently arouses serious doubts summed up by agricultural journalist Matt Mundell:

Those of us whose work takes us to the hill sheep country over Scotland, the Pennines and North Wales mountains become involved. Involvement turns to concern—concern at the ever-diminishing and changing quality of life that used to be in the hills, the emptying outbye cottages linked to the black spread of the trees in much of our sheep country. More schools closing, no toddlers in the glens, no mobile shops. Perhaps even a race of people vanishing.

THE HYPNOTIC EYE

4 THE SHEPHERD'S DOGS

There is no good flock without a good shepherd, and no good shepherd
without a good dog.

Motto of a French shepherds' society, adopted by the
International Sheep Dog Society.

For assessing the relationship between sheep, shepherd and dog,
this quotation cannot be bettered. Sheep farming on the lowlands
is difficult and time-consuming without dogs; on the hills it is
impossible. As Ralph Fleesh wrote in 1910:

Were twenty picked athletes sent out in the early morning to
accomplish the work of collecting and 'bughting', the chances are
that they would not reach the fold till the shades of night were
falling, and they might fail altogether. Sheep soon learn to outwit
man—they seldom challenge the prowess of a thoroughly trained
collie. It thus becomes clear what Hogg, for instance, points out,
that but for the collie sheep-farming would be an almost impossible
industry. And all the wages the faithful fellow asks are home-
grown meals, a straw bed, and a little kindness!

Each shepherd has at least two dogs, and probably four or five, to allow for injuries, bitches on heat or nursing pups, or partly-trained youngsters. Often the shepherd has kept the same strain for years, even though he may have changed his job several times. Inherited traits are brought out by the hill shepherd, who likes to train his own dog from puppyhood. Sensible strains of sheep dog breed a high proportion of pups with an eye for sheep. An example of this is the famous Mindrum strain bred by retired Northumbrian shepherd Bob Fraser. Any Border collie bearing the Mindrum prefix is expected to use its brains and work hard; it does not give in to an ill-tempered ewe, or turn tail when herding a really big mob. Some may be in the English national team, others just good honest work dogs.

The main difference between trials dogs and the rest is often just a question of extra polish. Some dogs never give of their best away from home surroundings; others are too severe on their sheep when keyed up in a trial's atmosphere. Many British dogs are now bred to International trials winners, but whether the general level working on the hill is higher than thirty years ago is a subject for fierce debate among sheepdog men. The International Sheep Dog Society records the breeding of all registered dogs, and excellent workers are also bred from strains of unregistered stock.

For the first few months, the pup is allowed as much freedom as modern traffic warrants. It is all the better for human companionship in the house, and to play with children may be the making rather than the undoing of a potential work dog.

As an example of the sagacity of working dogs, Ray Dent recalled: 'We had a black-and-white collie, Laddie, who could be set off here at Glenwhelt with either cattle or sheep, and he would walk behind them till they reached Swinhope Head two miles up the road. There our shepherd Jack Ross would be on the look-out, and open the gate for the beasts to turn in. Laddie would take the last one by the heel as a parting gesture, have a pat and a word from Jack, and trot back home.'

The most fashionable Border collie colours are black-and-white and tricolour. The tricolour is generally black, white and tan, but brown, white and tan is becoming more in evidence. Brown and white collies can be very handsome; one which comes to mind was used for lifting competitors' sheep after each run at the 1966 Chester International trials. With its rough coat and bushy tail it attracted as much attention as the top dogs.

Another colour making more impact is the merle. This has the white collie markings, which are on the face, chest, feet and tail tip, but the body colour is of broken black and grey. The impression is of torn patches of cloth. Very rare, though becoming sought after, is the red merle, an attractive mixture of brown and fawn.

Though most collies have dark brown eyes, the merle's are often bright blue, although some have eyes of different colours; many blue merles have one brown and one blue eye. A merle's blue eye may be mistakenly associated with blindness; this is not so, as the many excellent working merles prove. Shepherds do not always favour light colours for quite another reason; they know that a dog with too much white on its body is not easily seen by the sheep. Yet another colour variation is the mottle, which is a spotting of the legs and face and does not appear until the pup is three or four months old.

The word 'eye' has a distinct meaning in the sheep dog world. It refers to the amount of concentration on the sheep that the dog shows. Sheep are 'held' by the power of the dog's eye, and a dog in which this trait is well developed is termed 'strong-eyed'. In extreme cases it simply stares at the sheep without moving, becomes impervious to commands and is in general a great time-waster. However, the characteristic has been improved by selective breeding. A dog with little 'eye' is easier to break and to control. In close work, such as shedding, it induces less fear in the sheep, which tend to pack closely together for mutual safety against the 'strong eye'. Also, the relaxed dog is kinder to itself, using less nervous energy than the type that must watch every movement.

Though the young Border collie may evince its inborn herding instinct at an early age by rounding up ducks, poultry, or other dogs, it should not be introduced to sheep until it has speed to outpace them. A pup that cannot keep up with galloping sheep tends to develop bad habits, barking or becoming over-excited in its frustration.

The ideal training area is a small paddock, with seven or eight quiet sheep. Initial reaction from the young dog may be to chase the sheep or even bite them, but the skilled shepherd channels these actions into the desired course. If the pup makes no reaction at all, even the most skilled handler cannot improve it.

Voice commands are used initially. The youngster is accustomed to walking on leash, to lying down and to staying down while its handler is at ever greater distances. However, these early lessons may be forgotten in the excitement of seeing the sheep. Many hours of patient training go into the near-perfection exhibited by Border collies at the big sheep-dog trials.

Apart from Come Heel, four basic commands are used. The words used are irrelevant, provided they are consistent. The dog must learn to Stop or Lie Down. This it must do without question, and without creeping about once it is on its belly and no further command has been issued. It must learn to move to its right and to its left on command. A common phrase for the left hand is 'Come Bye', and for the right 'Away To Me' or 'Away Here'. The method of teaching this is to let the dog work the little group of quiet sheep, and as it moves of its own accord to its left, give it the Left command. The same is done for the Right, and gradually movement is associated with words, so that eventually the pup moves in the required direction when it hears the command.

The fourth basic command is Come On (To Your Sheep). This is the same whether the dog is bringing sheep towards the shepherd or driving them away; in either case it is walking up to the flock. Bringing the sheep towards its master is a natural instinct in most dogs; driving them away is contrary to its early lessons, and may take a lot of time to learn. Driving along a narrow lane is a help.

More refined commands are Go Back and Steady. The latter is obvious, but a dog that will Go Back on command is a great asset, especially to the hill shepherd. All the sheep in a certain area may have been gathered, but the shepherd knows that there are others out of sight, perhaps over a hilltop. He directs his dog accordingly, giving it either Left or Right signals as it goes if necessary. With a young dog, the shepherd sends it away only when he is certain that some sheep are there. An older dog may be sent to check whether any are left or not, and understands this, but a youngster might lose confidence and become confused, if no sheep were to be found when it thought it had obeyed its master.

In teaching a young dog to Go Back the first step is to break its concentration on the sheep it is already working, in itself the opposite of previous lessons and inherent instinct. A good way is to split a small flock worked close at hand into two, and after the dog has moved one lot a little way, it must be taught to leave them on command, and redirect its thoughts to the others.

Young dogs often show amazing wisdom. For instance, when Speed was eighteen months old, his shepherd came across a newly lambed shearling, which wouldn't stay close by its lamb. It watched the strange object from a distance, so Speed was sent to gather the shearling and reunite mother and offspring. He was left 'setting' or holding the ewe with his eye while the shepherd continued his lambing round. An hour and a half later, the young dog was still mounting guard, and mother and daughter were completely reconciled.

As training proceeds, whistles are superimposed on voice commands. A code of simple yet distinctive sounds is used, the long single blast being the universal Stop whistle. The Walk Up whistle may be two short blasts in quick succession while, for Right and Left, two varying long notes and four short ones may be used. At a sheep-dog trial, try to pick up the handler's commands by watching the dog's reaction to the different sounds. It is not as complicated or mysterious as at first appears.

Every hill shepherd from Exmoor to Sutherland has experienced a flock of sheep approaching, gathered out of sight by a dog on command, but no dog with them. This is a sure indication that the dog has found a sheep fast in a hole, and is standing guard until his master arrives to investigate. The amount of reasoning that goes into this move may be debatable; the safe result is not.

One dog, Moss, was out on the fell in late summer when a mist came down. Three-quarters of a mile away a group of perhaps a dozen sheep trotted out of the mist and immediately disappeared. Among them was a 'rough' ewe, one that retained her fleece while the rest were all newly clipped. The shepherd directed Moss with no more than the usual quiet signal to run round the sheep. Several minutes later the dog appeared, bringing the 'rough' ewe and her lamb, and no others. Was this sheer intelligence, or a case of dog and man so finely attuned to each other by hours spent on the open hills that the one knew the other's thoughts? Only the unclipped ewe was wanted, as she had obviously been missed in the 'clipping gather'.

Another dog that can bring out a particular sheep from a large flock is six year old Sue, a black-and-white bitch with cream face-markings belonging to Glencoyne shepherd John Maxwell. If one sheep, say a geld ewe among the in-lambers, is required, Sue will trace it through the flock without faltering. Her grand-dam was Meg, from Patterdale Hall, who died the day after whelping. Her pup, also called Meg, was reared by feeding at intervals through the night from a fountain-pen filler, and slept in a sock. After this upbringing she was anti-dog and pro-human throughout her fifteen years, during which she gave birth to Sue.

Some collies have quite incredible noses. One lambing time above the 2,000ft contour, a ewe, some three hundred yards away, was going round in a big circle, 'blaring' at the top of her voice. Obviously she had a lamb missing, but where? Among the haggs and peat bogs it might be anywhere. The collie 'on duty', Rock, immediately set off in a bee line for a hole in the peat two hundred yards away. Following the trail, the shepherd rescued a lamb, black with peat. He carried it towards the ewe, which didn't recognise her offspring because of the peat. Off she galloped for a mile and a half, with the lamb still following. Rock and his master then caught the lamb, washed it in the nearest burn, and it was finally accepted by its unruly dam.

An old dog, Ken, had a wonderful nose but a bad temper. Crossing a high fell one day he put his hackles up and his nose down, and set off towards some bare peat 200yd away. In the peat were the local keeper's unmistakable tracks—size elevens— and later enquiries ascertained that he had passed that way two days before. Yet Ken had scented the stale track from a considerable distance.

If Moss, walking at heel through a field full of lambing ewes, suddenly started to weave behind a particular sheep, the shepherds could guarantee that that ewe would have lambed by the next round. This 'lambing smell' was so distinctive to Moss that if a newly dropped lamb had become parted from its mother among a packet of sheep, the dog could scent to which ewe it belonged. When gathering near the boundary, he could distinguish between his own and a neighbour's sheep. He would allow the latter to drop back, but never lost his own.

When Moss was twelve years old, he was retired, and became fat, heavy and deaf. One morning as his shepherd was setting off with tractor and trailer, Moss came up, looked at him and begged to be allowed to accompany him. Stopping near a large stretch of fell, the shepherd found a few ewes that needed driving away, and thought Moss would love to do his old job, so he set the dog behind them, a fifty yards' drive being planned. Moss thought differently. Right round the boundary of 600 acres he went, and being deaf couldn't hear anything of his handler's shouts and whistles. Because of his weight the old dog could only paddle slowly round, and the shepherd's exasperation grew as time passed. Eventually Moss arrived with a large packet of ewes, and never did a dog look so happy, smiling and pleased with himself.

Most sheep dogs have their own peculiar traits. Rock, for example, was directed to gather sheep round the bottom end of a felled wood, out of sight. When he returned he dropped a hare at his master's feet, having picked it up from its form and carried on with his work. Unfortunately the shepherd did not see the hare till it was released from the dog's mouth, when it immediately ran off! Rock's mouth was so gentle that he would let young lambs—the most awkward things to drive—run into his mouth and turn them without hurting any. Another dog, Glen, wouldn't turn lambs till clipping time. He couldn't be bothered with the little, smooth things, yet as soon as the ewes lost their jackets and were smooth too, he was all right.

Though hill sheep graze on hefts, or particular areas, they must be brought together for occasions like dipping, shearing, vaccinating, or selection for sale or tupping. This collecting is called 'gathering', (see Glossary). The worst weather for gathering is low cloud or hill mist. To attempt it under such conditions is hopeless, the most disheartening combination being a clear start followed by sudden and unexpected fog, entailing abandonment of the gather. The other real test comes when rain and driving wind make voice or whistle inaudible 50yd away. The dog then relies on hand directions, rather frowned on in the trials field.

All sheep dog trials are great fun for the shepherd and his family. Occasionally they become a mania, the dog taking precedence over the sheep, but in general they are a stimulus to better shepherding and dog breeding. No weather conditions, except fog, cause abandonment, and from early morning till evening competitors arrive with their collies, make themselves known to the secretary, and take their place at the starting post.

The line of cars drawn up at the ringside contains collies of all colours, ages and temperaments. Some are tethered to the hedge, others wander near the course while the doings on the field are the focal point of interest for a number of keen-eyed canine critics. For each competitor a small group of sheep is let out at the far end of the course between a quarter and half a mile away. First stage is the Outrun, in which the dog is directed from the handler's post to run in a widening arc behind the sheep. Points are deducted for faults, such as going too far or not far enough, or for an irregular outrun needing redirection signals. The next stage is the Lift, the few seconds in which the collie stops behind its sheep and they make their first acquaintance of it. This first impression is vital; a rough or jerky dog upsets the flock and causes more work later.

From the Lift the collie fetches the sheep towards its handler in as straight a line as possible, passing between two hurdles en route. Driving its flock behind its handler who has remained at the starting post, the dog is then directed on the Triangular Drive, aiming for two more sets of hurdles through which the sheep must be safely steered.

The third leg of the Triangular Drive brings the flock into the Shedding Ring, a sawdust-marked area within which two or more marked sheep are separated from the rest. In the case of the Single, just one sheep is shed off, and these manoeuvres are carried out in midfield without wall or pen for assistance. The idea is to emulate working conditions as closely as possible.

Penning is the final stage. The pen is a small enclosure with a gate 7ft wide, or no gate at all in the case of doubles (two dogs). The handler holds a rope at the end of the gate and assists, but must not touch the sheep. Penning calls for a combination of steadiness and quickness on the dog's part, as too much pressure results in sheep making a mad dash round and round, while exact positioning is necessary to steer the flock home.

In Brace or Doubles running, two dogs are directed upfield, one on either hand, and between them they work the sheep on the same course. The difference is in the shedding ring, when the flock is split in half, and one dog pens its own sheep and then guards them in the gateless pen. Sometimes the temptation to help its pal is overwhelming, and the first collie slips away from its post of duty, giving the inmates chance to slip out. Doubling has the fascination of all team work, and is a prime favourite with spectators. Yet away from the glamour of the big trial, these same dogs will be working over the quiet fells with only the curlew and golden plover for company.

The International outrun, the largest competitive one, is half a mile, and it is by no means easy to find a field of that size with adequate spectator facilities. In practice, however, distances are often much greater. To gather one side of Swinhope, Ian Dent's dog, Fly, takes a three miles' outrun before returning. In doing so she must work in and out of the gill heads, as the small steep-sided cuts near the start of the many burns are termed.

Ian had another bitch, Lassie, that suddenly left him one evening in lambing time, popping through the fence and away half a mile uphill into a thick, black mist. Ian, knowing that the dog had gone with a purpose, followed into the dark, eerie thickness and came to a ewe having difficulty in lambing, with only the lamb's head hanging out. But for the dog that lamb would certainly have died.

All shepherds know the value of their dogs. For the unenlightened general public, James Hogg, the shepherd–poet from beautiful Ettrick, summed up the worth of the Border collie with these words:

Without the shepherd's dog, the whole of the open mountainous land in Scotland would not be worth sixpence. It would require more hands to manage the stock of sheep, gather them in from the hills, force them into houses and folds and drive them to market than the profits of a whole stock would be capable of maintaining.

D.M. & E.M. ALDERSON
77

5 HIS TOOLS

Speakers at shepherds' dinners have been known to say that the only things needed by the profession are a dog, a crook and a pair of boots. This may be a slight over-simplification, but basically the hill shepherd works effectively with the very minimum of equipment. Despite the impact of science on controlling sheep diseases, the tools used by the man on the ground are uncomplicated.

Making the crook is described in Chapter 10. Using it is also an art, perfection in which takes hours of accomplishment. The crook is an extension of the shepherd's arm. It is of two main types, neck crook or leg crook. The former is shaped to allow the narrowest part of the sheep's neck, just below its head, to enter. The crook is then twisted under the sheep's chin and the animal cannot escape.

This sounds easy, and among a flock of ewes in a yard it is. Our shepherd is more than likely working on a slope with no walls or pens at hand. He sees a particular sheep he needs to catch; the reason could well be lameness, a sign of ill health, or to treat her young lamb. The dog is directed round a group with the wanted animal among it, and drives it close to its master. This may be at full gallop with the slope lending speed to the flight, but by outstretched arm and extended crook the ewe is caught by the neck and brought to a halt in a spinning bundle of wool.

This art, properly accomplished, is as satisfactory as scoring four runs off the middle of the bat, with no danger of being given 'out'! If the sheep is missed the collie merely repeats the process, but a professional shepherd does not often miss. Round come the sheep, out goes the arm at full stretch, and the sheep's neck runs along the top of the shank and into the cunningly shaped mouth of the crook; she twists with her own momentum and lies helpless.

A Welsh or Blackface ewe charging at full pelt downhill creates some force. A weak crook would snap under the impact. A horn head will not break; coming from the sheep itself, it is fully able to withstand any power the animal can apply to it. The danger point is where head and shank come together, and this join is the real test of the stick dresser's art. Nor does a crook made from the root of a hazel give way. The twists and knots of the rootstock stand firm against the hurtling sheep, but a head carved from the branch may split if the grain runs across rather than along it.

Smaller, and more of a 'Sunday afternoon' crook, is the 'lamb stick'. On a lighter scale, its neck is just wide enough to take the head of a young lamb which is as evasive as an eel when only a few days old. A short-handled light crook is useful for working among penned sheep, enabling ready examination to be made, while for most pen work the leg crook or leg cleat is best.

With head of wood or horn, this crook swells out slightly from a mouth the width of an old penny. Its purpose is to catch the sheep just above the hock, when the crook is drawn backwards to the handler. A mob of tightly packed sheep turn their heads constantly from the shepherd, each animal trying to escape

Plate 8 The shepherd's crook comes into play when counting sheep. Accurate checks must be made whenever the flock is gathered, for the next opportunity may be weeks or months hence. The trick is to funnel the flock through a narrow gateway, allowing only a thin stream of sheep through. Very rapid counting and undivided concentration is called for; the speed at which an experienced shepherd counts sheep is an eye-opener to people used to dealing with figures in offices. Here Herdwicks come down the hill for the low ground lambing pastures at Glencoyne Farm, Ullswater, Lake District

among its fellows. In such circumstances it is by no means easy to force a neck crook over the desired head, and even then the animal must be turned against the weight of the flock before being separated. The leg crook properly used enables the shepherd to slip his stick round the leg from the rear, and draw the sheep backwards towards him.

Sticks may also be of aluminium or stainless steel. They do not break, and if left out in the rain for a fortnight do not lose their shape as does a purchased steam-bent hazel or ash. They are practical, cheap, and abominable! They are no comfort in the hand, whereas a well balanced wood or horn crook becomes an old friend without which no journey to the hill, or even across the yard, is contemplated. His crook to the hill shepherd is like a horse to the American cowboy, who would walk half a mile to catch one particular horse to ride a hundred yards.

A pair of good boots are a shepherding essential. They must be watertight, with well nailed soles to prevent slipping on slopes. They are not cheap. The 'sprung sole', hand-made on a Scotch last, cost over £30 a pair. Their curved soles are used with a rocking action that carries the wearer on at each stride, and the boots are not as heavy on the feet as they feel in the hand.

Care of boots is essential if they are to last. Shoe polish over the stitches, and vegetable oil or dubbin elsewhere, preserves them. The laces should be tightened without twisting. Made from best leather, they should last the life of the boots, says Cumbrian bootmaker Mr Otway, from Ambleside. He still twists his own threads for repairs, and criticises the farming community for the neglect they give to an increasingly expensive item of equipment.

Every shepherd carries a penknife and lengths of baler twine. The former is multi-purpose; it may pare an overgrown hoof one minute, trim a sharp-pointed horn the next, and then cut a length of the string to tie a sheep by all four legs until it is collected later on the round. Some shepherds go through dozens of knives in a lifetime, but one Kirkby Stephen (Cumbria) upland farmer solves the problem by carrying one so big and heavy that he immediately knows when it is missing!

The local district nurse was astonished when she saw the preparations for lambing at Glenwhelt every year. Familiar though she was with human midwifery, the range of lambing oils, string and appliances came as a surprise. All are carried in hessian lambing bags, other bags being prepared for carrying lambs. Strong twine is needed for pulling back limbs inside the ewe's body, and very fine string is used when the mother bites off the umbilical cord too close to the body of her newly born youngster. The wound is dressed and secured. Lambing hooks also come in handy for complicated births, though it is rare for a hill shepherd to have more than twins to deal with. He misses those rare mix-ups which occur in fertile lowland flocks when three or four heads, three or four tails and six or eight legs seem apparently inextricably locked.

A piece of towelling is standard. The shepherd may wash his hands a score of times, though some regret the present lack of lambing oils which were wiped liberally on hand and arm, and which washed off in the nearest burn to leave clean flesh. Marking fluid is carried in a screw-top jar, for each lamb is given the same 'pop mark' as its dam as soon after birth as possible. Baler twine is always carried; one modern benefit is red-coloured twine, a length of which tied round a ewe's horn forms an easy and distant identification if the shepherd wishes to check that she has taken to her lamb, thus avoiding close approach and further disturbance.

As emergency reviver, hot coffee and sugar are difficult to beat; they quickly reach the animal's loins, the most exposed part. Shepherds generally believe in walking as light as possible, and in some years their array of tools is scarcely needed. Other seasons bring a crop of awkward lambings, the reason being seldom apparent.

A number of sets of waterproofs is needed, however, for whatever the weather the lambing round must go on; indeed, the worse the weather, the more time must be spent outside. Some hill men find a hood essential; water trickling down the back of the neck is cramping and chilling, making the walker hunch up instead of looking about him.

When science reached the sheep fold, shepherds took to the hypodermic syringe like ewes to hay in a storm. 'Give her the needle' became a sovereign remedy for every ill. Modern routine vaccinations against such soil-borne diseases as blackleg, braxy or pulpy kidney (see Glossary) are now well understood, and the multiple five-in-one or seven-in-one give protection against fatal illnesses that bedevilled our grandfathers. Lambs acquire a certain immunity while in the womb, so one set of injections is given during pregnancy. The other is usually in autumn, both being timed to coincide with flock gathering for some other purpose, such as dipping or worm drenching.

Though hill sheep may not be subject to so many stomach worms as intensively grazed lowland sheep, they nevertheless tend to frequent certain areas, old barns or the banks of streams. Here worms multiply with amazing rapidity, so that the drenching gun and its use are very much part of modern hill shepherding. Various types of worming gun are on the market. One squeeze of the trigger administers a pre-set dose down the unwilling throat. At least, that is the theory, but guns may block or become airlocked, and if a sheep gets the wrong dose she may as well have none at all. There is still much to be said for the small dosing bottle, filled for each sheep from a row of squeezer bottles strategically placed on the wall of the working race.

Shepherds have their own fads and fancies, and large-scale Northumberland farmer Frank Walton allows each of his shepherds to use the worming brand of their choice. The discounts of bulk buying may be lost, but Frank Walton understands the impossibility of attempting to persuade his independently minded 'herds to use something in which they do not believe. The quality of those Blackface and Cheviot flocks grazing the Redesdale firing range and towards Cheviot itself vindicates the policy.

A pair of hand shears is still necessary equipment, despite the development of mechanical shearing. On remote hill farms, power may not be available, or the outlying hirsels so distant from the homestead that it is impractical to bring the flock home. A sheep may become soiled under its tail and require clipping out. Since the ban on dieldrin (through fears of its persistence in wildlife and eventually in humans) more and more sheep are 'struck' with the blow fly. Clipping affected parts is among shepherding's most unpleasant jobs.

Horned rams may be fitted with devices to encourage the horn to grow the correct shape. Some tend to grow too close to the face, others too wide apart, and a quick traditional way was to boil a turnip, slide it onto the horn and let the heat soften the hard cover. The horn could then be bent to the desired shape.

Spring comes late to the hills, but when it arrives it brings a load of troubles. The shepherd then sets off with a bottle of

Plate 9 Use of shearing stools in Lakeland. A colour mark is being applied to the newly shorn sheep on the right. Stools tend to be used for small breeds like Welsh and Herdwicks which entail more bending. The right-hand pillar is a fine example of stone masonry for which many upland districts are famous

calcium-magnesium mixture to inject into any ewe that goes down with 'staggers'—a temporary absence of the required mineral in the blood stream—when the animal lies down, becomes unconscious and dies unless the syringe is used immediately. If caught in time, the cure is so quick and complete as to appear miraculous.

If a flush of growth comes in June, the shepherd carries a pair of shears on his back, to trim any sheep made dirty as the lush grass passes through her. Lambs are ear-marked with special pliers in which the appropriate punch hole or slit is set. Another spring job is castration of males not wanted for breeding, although a final selection is not made till later in the year. The painless and bloodless rubber-ring method is most common, the shepherd using a pair of castrating pliers to stretch the ring over the testicles.

A mass of equipment is assembled for shearing. The barn is

thoroughly cleaned, top, sides and floor all receiving attention. A catching pen with a swing gate is set up, clipping boards laid in place and the machines hung above them at exactly the correct height. A table for wrapping fleeces is set up and scrubbed, and another table for odd items of equipment put in a suitable place. On it go saws (for overgrown horn), foot-rot stuff, one or two types of worm drench, hand shears for dirty wool and a sharpening stone, besides big shears for hard hoof. Oil for the machines and spare combs and cutters are set out. These combs and cutters are the sharp edges of the shearing-machine head. Sharpening them is a skilled job, and at least a day's supply must be stocked.

On some farms nothing else is done on shearing day. This applies more on the lowlands, where it is comparatively easy to bring in the flock for handling and inspection. On hill farms the sheep are gathered only on specific occasions—for tupping, dipping, lambing, shearing and weaning. Opportunity must thus be taken to attend to any lameness or lack of condition, for the chance may not recur for several weeks or months.

Those distinctive red, blue or black fleece marks have obviously vanished with the wool on shearing day. They must be replaced as soon as the sheep is shorn, and a marking iron and can of paint stands on the floor. Lambs are not shorn till their second year (except among a very few sheep in the South West). A tin of white or red lead paint is also kept handy, to mark on the horn any ewe faulty in her udder. This is an easily recognisable signal that that ewe is not to be sold as sound with the regular drafts. She is fattened or, if kept, that flash of paint is a signal that if she lambs twins (she usually does!), one must be taken off.

Medical oils are placed ready to apply to cuts, but these should seldom be needed. Contractors' gangs working at high speed usually leave more cuts than shepherds who shear the sheep they deal with daily. Ray Dent, for instance, employs no extra help for shearing. 'We don't set off to race or kill ourselves; there is always another day looking at us', he says. 'Then we have a two-mile walk from folds to fell at the end of the day, so 140 sheep a day for two men is about average. We have also carried out any examinations and treatments necessary.'

'Speaning' or weaning time is in August, when lambs are taken off their mothers by running the flock in single file through a drafting gate, lambs one way, ewes the other. In Australia four-way drafting may be practised, and any unfortunate man who sheds one of the non-stop flow the wrong way receives scant sympathy.

Weaning is often combined with a summer dip against the blow fly, in contrast with autumn and winter dips which give protection against sheep scab, and various other mites and parasites. Tools used include a pushing stick to duck heads under, a brush, shovel and a marker stick to see that the correct amount of replacement dip is added. A waterproof apron and old clothes are other essentials, though some of the new 'walk-through' baths are considerably easier on both sheep and shepherds than the type where each animal must be dropped bodily into the strong-smelling liquid.

One of the hill shepherd's best general purpose tools is a crate on the back of the tractor. It has high sides and a door and, when lowered by the tractor hydraulic, a ewe and lamb may be driven straight into it. A small crate, 18in by 12in, takes three lambs and prevents them being trampled by the ewes. Dogs are transported to and from the fell in this crate, and enjoy their rides. Such a simple contraption saves hours of time and effort, for to walk a ewe with newly born twins right across an allotment is a very slow process. Bales of hay are taken to the exact point required, and farm tools like spades and crowbars carried safely.

EXMOOR HORN RAM.
DM & EM. ALDERSON 1977

6 SHEEP MARKS

'Found, Blackface ewe, back-nip near lug; four saw marks, near horn, red pop mark near flank, black stripe off shoulder', reads an advertisement in *Scottish Farmer*. The wording, with variations, is repeated wherever hill sheep are kept, and, to the experienced shepherd, immediately proves the animal's identification.

Many hill sheep are kept on unfenced common ground. In mountainous areas, such as the Scottish Highlands, fencing is quite impractical. If fences could be set into the stony soil, the next series of winter storms would only bring them down. So each farmer must be able to identify his own sheep in a manner far more detailed than on the plains.

Marks are of three types—ear, horn and fleece. The ear-mark is made with a special pair of pliers soon after birth, for a little hole becomes a big one as the sheep grows. In some cases the tip of the ear is cut straight across, or 'stowed'. A lamb's ear is thin and fine, and such marking is accompanied by only momentary discomfort and a few drops of blood. 'Upperbit' and 'under-

FLY IN SHEEP.

Maggots on **Sheep** are **speedily destroyed,** twisting and turning out of the wool almost immediately, by using

Cuff's Fly, Scab, & Mange Oil,

which also prevents the Fly striking again, and heals the wound quickly.

Price 1/8 and 3/- per Bottle.

FOOT-ROT

is cheaply and effectually cured by using

Cuff's Foot-Rot Powder,

Price 1/- and 2/6 per Packet.

CUFF'S FARMER'S FRIEND

THE reliable and well tried remedy for preventing and curing **Scour,** assisting Cleansing, preventing Milk Fever, etc., has often cured cases which seemed hopeless.

Price 1/- and 2/6 per Bottle.

Original Driffield Oils

Prevent gangrene after lambing or calving.   Heal Cuts, Wounds, Bruises, Strains, Swellings, etc.

Price 1/6, 2/6 and 5/- per Bottle.

PREPARED BY

J. H. CUFF & SONS, Cattle Market, LONDON, N.
And sold by Chemists throughout the United Kingdom.

Two pages from the *Wensleydale and Swaledale Almanack,* 1911

Register of Sheep Marks.

On the following pages full particulars are given of the horn-burns, tar, rud and ear marks used in the undermentioned flocks.

Any persons finding stray sheep with the marks herein given are requested to communicate with the owners, who will pay postage and all other expenses.

The charge for insertion of marks is ½d. per word; minimum 6d. Block of sheep, with marks in proper position, 1/6 extra.

HORNBURN.	OWNER.	TAR MARKS.	OTHER MARKS.
EA near	Edward Allen, Mile House, Hawes	Ewes: ● near shoulder	Lambs same tar mark and slit near ear
EA near	E. Allen, Gayle, Hawes	A near side	rud on shoulder Lambs: EA near side
EF	E. A. Fawcett, Birk Rigg, Hawes	O both sides	
EH near	E. Hall, Marsett, Askrigg	rud near fore shank	Lambs: ● far ribs
FL	F. Lambert, Brough Hill, Bainbridge	FL near side, stroke down far shoulder	
FP	F. Paley, Countersett, Bainbridge, Askrigg	P near loin O far shoulder	Lambs: O far shoulder
GC near	Geo. Coates, Raydale House, Askrigg	⊖ near ribs Forelands Sheep: blue far ribs	winter rud far ribs Lambs: ● on bottom of each hind shank
GM	Geo. Metcalfe, Low Blean, Askrigg	when tarred ● far loins	rud near hind shank
GP	Geo. Preston, Raygill, Hawes	P near side	

bit' are terms for notches in the top and bottom of the ear respectively. Further permutations are gained by square, round or triangular notches, while Dr Jakobsen in 1909 described the Shetland term for two 'bits' on opposite sides as 'gongbit'.

A punch hole is given in the near (left-hand, when facing in the same direction as the sheep) or far ear, or both. One disadvantage of the punch is that it may interfere with the metal or plastic ear-tag which has both flock name and that sheep's individual number. Ear-tags are convenient and make individual recording much easier, but a small proportion of them do tear out. The older methods are safer, but even they are known to have been tampered with, and altered to suit a thief. Sheep stealing is rightly regarded as a serious crime, for it destroys the community spirit of a hill district when no one knows the destiny of missing sheep.

Different upland areas of Britain have their own registers of sheep marks. In 1819 four North Westmorland farmers began what has become an invaluable service for northern sheep farmers. They published the very first edition of *The Shepherd's Guide*; or a delineation of the wool- and ear-marks on the different stocks of sheep on the East Fells.

Markings of the sheep of 233 farmers whose animals grazed the common land and fell of East Cumberland and North Westmorland were gathered together and published in the *Guide* and, ever since, the area's farmers have had a reliable system of claiming strayed sheep. The *Guide* has been revised every ten years, on the latest occasion by septuagenarian Harry Wales, of Lownthwaite, Milburn, who had over 2,000 markings to check.

In the original issue we read:

It is well known to every proprietor of sheep how apt they are to stray from their owners, and consequently, either from not knowing the proper owner or from neglect, or a worse cause (the fraudulent intent of the discoverer) are often entirely lost to him.

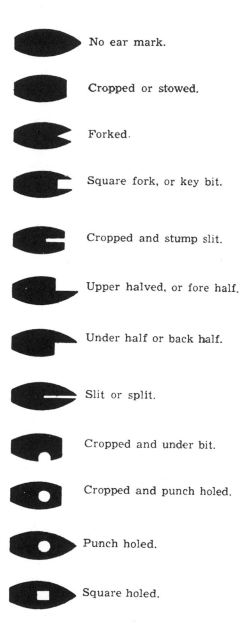

No ear mark.

Cropped or stowed.

Forked.

Square fork, or key bit.

Cropped and stump slit.

Upper halved, or fore half.

Under half or back half.

Slit or split.

Cropped and under bit.

Cropped and punch holed.

Punch holed.

Square holed.

Entry 122 could apply almost exactly today: 'Isaac Slack, Skirwith Hall. Tar mark, a pop on the far shoulder, a gelder horn on the near hook; ear mark, a little off each ear; horn burn IS and a figure.' Only the name has changed. The present farmer is Mr Guy Slack, a direct descendant.

Two shepherds' meets a year were planned. According to the rules, all stray or neighbouring sheep had to be at each meeting place by noon, 'and those sheep whose marks are not in the book, to be drawn into a fold by themselves, for the inspection of all present at the Meeting. Members who infringe this rule be fined sixpence each.'

At the end of the *Shepherds' Guide*, all horn burns for the area are listed alphabetically. Here is a typical selection from the 1920s edition:

HORN BURN	EAR MARK	NAME	PLACE	PAGE
IC near	Hole near	Wm. Cowper-thwaite	Roun-thwaite	294
ID near	Far underbit	Willis Dunning	Tebay	295
IW far	Bit off end far	Isaac Wilson	Old Tebay	303
IB near	Forked near far underbit	Isabella Bryning	Newbiggin-on-Lune	305
ID near	Far underbit	Robt. Dunning	Tebay	307
IC near	Near cropped	Thos. Porter	Tebay	308

Ears may be simply slit with a knife, or a narrow cut shaved off one curve—'strae-draw' as the Shetlanders termed it. Their description of a slit down from a square top was 'rit i de stoo'.

The horn burn is as permanent as the horn itself and, though horns do occasionally break off, this seldom occurs until old age when the sheep is well known or has left the hill. One seldom sees a broken horn in a thriving hill flock. In breeds like the Cheviot, Derbyshire Gritstone or females of Welsh and Herdwicks, there is no horn to burn, and reliance must be placed on other methods.

Upper shear bit, or fore shear slipe.

Under bit, shear bit, half moon, fold bit, or back bit.

Under key bit, or back key bit.

Upper bit, or fore bit.

Upper key bit or fore key bit.

Dove tail bit.

Club ace bit.

Spoon shanked.

Under slit.

Diamond punch hole.

Oval punch hole.

Usually in the form of initials, these horn burns identify the owner rather than the particular heft of each sheep. They are handed down from one occupier to the next, and may bear no relation to the initials of the present breeders.

When I farmed in Bilsdale, North Yorkshire, my sheep were branded JH, the initials of the father of the previous tenant. Branding irons are of two sorts. Ready-made letters are clipped into the iron head, or the letters are cut direct into the head by the blacksmith. The latter process is more expensive but longer lasting, and a more even heat is obtained. A wood fire is best for branding irons; small clinkers are formed from coal fires, and the heat from a cooking stove is too severe for the metal.

This operation is, of course, painless, there being no nerves in the outer horn. A specially important brand is the 'crown' used by Swaledale Sheep Breeders' Association as a mark that the sheep has been approved by the inspecting panel and is of correct breed type. Many thousands of Swaledale rams and shearling ewes are crowned annually, and replacement of crowning irons is a regular expense borne by the different districts of the Association.

Horn burning at Ray Dent's Glenwhelt takes place after weaning. Dam and offspring must be separated for eight days at least, otherwise the milk flow will not have dried up. After that time, mother and daughter may still recognise one another, but no suckling takes place and the ewe lamb returns to the fell. Not only is she out of the way, but she learns the boundaries of the heft, which she will one day pass on to her own offspring.

In this same August period those gimmer hoggs retained for breeding receive a red fleece mark. After clipping, both ewes and lambs received a black mark, and this change to red is a check on any stragglers. If a hogg is found in September on the far slopes of Chapel Fell marked in black, it is obvious that she was one that missed the count in August.

A sheep may well carry three different fleece marks. A red V down the near buttock might identify the flock, while a black or blue stripe down the far shoulder indicates the particular heft.

At tupping, a 'pop mark',—a dot of a particular colour in a particular place—indicates the ram with which that ewe ran. As soon as her lamb is born, it receives an identical pop mark, so that its sire is known.

These marks fade with time, and are designed to scour out when the wool is processed. A sheep may rub its new mark against a peat hagg and obliterate it in that way. Tar marks are now banned by the British Wool Marketing Board, as the tar is impossible to scour out, and any fleeces so presented are severely penalised. Red, green, blue and black are the shepherd's favourite marking colours. All can be seen at a distance, red being usually the clearest in dull visibility.

Red on the ram's chest or on the ewe's tail top is different altogether. The ram's chest is smeared to see how many ewes he has mated, the colours being changed week by week so that it is known which ewes will lamb within certain dates.

6—ANN BAINBRIDGE, Cote House. Horn burn WB on the near horn ; tar mark, ewes and shearlings W on the near side and B on the far side, lambs tarred on the neck ; wedders keeled on shoulder.

7—WILLIAM GRAHAM, Healey Field. Horn burn EG on the near and P on the far horn; tar mark, G on the near and P on the far side, two-year-old sheep G on the near side, P far buttock ; Autumn tar mark, blot on near hook, ear mark uncertain ; Low fell sheep EG on near horn ; tar mark, G on near side.

8—JANE ELLIOT, Dean Howl. Horn burn TE near horn, P far horn ; tar mark, E near side, P far side ; Autumn, blot far buttock.

A page from *The Shepherd's Guide* from the 1920s

66

GUINNS' for a crosse ewe.

7 LAMBING AND CLIPPING

Young lambs frisking in warm spring sunshine have been pictured since art and shepherding began. On the hills the reality is only too often very different. A slow dawn shows the surrounding tops in dim outline against solid cloud from which falls an unceasing downpour. The shepherd leaves his warm bed as unwillingly as any commuter, not daring to lie for a few minutes listening to the steady patter on the windowpanes lest sleep once more overtakes a body weary from long days of tramping in heavy oilskins. A quick cup of tea, cornflakes, or porridge prepared the night before, fortify him for the first round till his fullscale breakfast around 9 o'clock, three hours hence.

Seeking weatherproof clothes, hopefully dried overnight, he dons boots or wellingtons and steps into the empty yard. Opening a loose box door he calls a single name—Bess—and the rest of the dogs kennelled there make no objection to further time on their straw bed, for they too are becoming tired and worn by the weather. Through the rutted gateway puddles pass the man and his dog, the collie turning her head against the beating rain, her master with lambing bag on back, crook in hand, hooded and oilskinned like a spaceman.

'A shepherd's job in lambing time is to search for things he hopes he won't find.' He smiles despite the sodden ground hindering every step, despite the constant splash of small puddles drenching his boots. The weather's effect doubles in intensity through its effect on the young lambs as they stand, backs arched like staples, under a wall or in the lee of a tuft of rushes, so wet and miserable that the shepherd can scarcely tell which are merely drenched and which are gaining insufficient milk. 'There's a new one', he says, as a small snowhite body stands in ever-surprising contrast to the grey of its dam and the dull green of its surroundings.

'Here Bess.' His dog slips silently away, so attuned to her master's wish that she knows that this, of all the scores of sheep in the field, is the required one. She drives it towards the shepherd, its red-navelled lamb staggering at foot. With a few quick steps and outflung arm he extends his crook and takes the ewe by the neck. Round she swerves until caught by flank wool and then horn, then set on her rump and with a swift movement any loose wool is swept from the udder, each teat tested for sufficient milk supply and checked that the lamb has indeed sucked. A flaccid teat indicates that the youngster has had its fill; a proud one is cause for future watchfulness, and the lamb may be caught and held with mouth encircling the teat if it is several hours old and should by now have sucked on its own. In any event it is caught, and marked in the identical place and colour as its dam. Its sire is thus recorded, and if it is an outstanding tup lamb from a good strain of ewe, it may receive a special mark—often on tail top—to mark it out for future selection as a potential flock sire for home use or sale.

Not every ewe is caught. If all looks in order, she is allowed to trot away quietly after her lamb has been marked. The shepherd's eye is trained by experience and tuition as surely as a successful racehorse trainer knows the precise moment his charge reaches peak condition.

When a lamb is obviously thriving and trotting strongly with its dam, they are moved from the lambing field into an adjacent area. This steadily reduces the number of mouths and the number to be 'looked', and is usually done once during the day. If ewe and lamb happen to be handy for one of the gates or 'smouts', the opportunity is taken to push them through there and then; on the next round they may be half a mile away.

On into the rain go man and dog, noting another new birth, then dropping down into a steep-sided gulley where rushes a torrent of water dangerous to young lambs. In the natural shelter thus formed, a bevy of sheep has inevitably collected. They move before the dog, but the shepherd takes the greatest care that small lambs cross safely or, better still, are driven uphill to safer if more exposed pastures. An annual toll is taken of lambs born besides such rushes of water; their first staggering movements of life become their last. Sometimes well grown lambs several days old are victims; they feel they can jump after their dams and over-estimate their ability.

For a hundred yards nothing of note is found. Ewes are grazing steadily in the downpour, none having lambs at foot or showing signs of lambing. Then on a headland the shepherd sees a lamb snuggled against the wall fifty yards below. Its dam is grazing near. He looks steadily, years of observation which have culminated almost in instinct raising a small doubt. Through the rain he sees that the lamb is sheltering behind a stake driven close to the wall. He whistles to move the ewe, and the lamb, though responding, remains fast.

Down the steep slope the man half runs, half slithers, till he comes to the place and finds that the lamb's head is pushed through the mesh wire strengthening the wall at that point. It might have had the sense to draw back; or it might have panicked to become inextricably caught, and dead before the next round. The gap between stake and wall is blocked with two stones, making one less hazard on the extensive lambing ground.

Further on two sets of twins are watched. Born the previous day, they will be pastured apart from the singles, but at this stage the shepherd is concerned that each is receiving equal milk and maternal care. Once a twin becomes weak, it trails behind its fellow, gets still less of the life-giving milk and falls an easy prey to fox or carrion crow. If a ewe at home has no lamb, one of these twins is taken to mother on, the stronger being chosen so that it withstands initial non-acceptance the better.

Here is an area of broken ground. Deep gullies and bare rocks split its surface, gutters run in the rain where none exist in a dry time, twists in its topography give shelter from every wind. Though such places are an advantage and give more shelter than the open moorland plateaux of Northumberland, they cause extra work. Only by surmounting every hillock is the shepherd able to discern what is happening over the other side, so ten valuable minutes are spent in checking quite a small space.

Suddenly the rain ceases. The wind drops, the air becomes clear and sheet-like wisps of mist hang over the hill ends. Birds sing. The plover or peewit calls its mate, a curlew's warbling note sounds across the valley as the big bird glides and touches earth, white undersides of wings meeting momentarily above its back. Ahead a number of golden plover make sweet, eerie music and the sun comes out. Clammy oilskins become heat traps, rain beads wiped from the shepherd's forehead threaten to change into sweat, and small lambs move from their shelter.

In fine weather, and with no disasters among the flock, shepherding seems the finest possible life. We step forward with new heart, catching a shearling that lambed during the night but whose lamb has that narrow, 'clammed' appearance of an empty stomach. The first-time mother has enough milk, but only just, and if the next night threatens stormy she and her lamb will join the ever-changing throughput of sheep needing special attention in the few buildings grouped round the weather-beaten dwelling-house.

The turn home begins, but with the sun comes tragedy. A small lamb with eyes pecked out is found, dead and deserted by its mother who moves undecidedly over the next mound. Carrion crows have been about their deadly work, attacking this lamb perhaps in the very act of being born. These birds may serve some purpose in tidying the countryside of corpses, but at this moment a hatred black as the bird itself fills the hearts of those finding a defenceless lamb killed by them.

Round the next wall corner the shepherd finds what he expects, a two-shear (three year old) sheep with an exceptionally fine gimmer lamb. 'That ewe won at last year's spring show, and she has a lamb every bit as good as herself', he says proudly. 'The ewe's mother is at home. She's eleven this time, reared twins for the past three years, never missed having a lamb and is in lamb again this year. She's failing now, but a stock like that is worth carrying on.'

The gimmer lamb is by a home-bred tup out of another well tried ewe. 'Her mother was by a Low Burns tup, her grandmother by a Gold Cleugh tup, and great-grandmother by a Wolf Hills tup. We can trace back these good families on both male and female side.'

A ewe with big, bouncing lamb ambles ahead sixty yards away. 'One of the "artful dodgers" ', says the shepherd ruefully. 'There are two or three ewes that hang around the bothy and spinney in mid-field, and when I go up one side with my dog they nip round the other. When their lambs are newly born they refuse to be stirred at all, then when the lambs can travel they keep their distance. Only way to get them is to put a dog up either side, but I haven't had time yet. When moving ewes and lambs from the lambing field, it is policy to take the easy ones rather than concentrate on the awkward sheep, otherwise you find yourself at the gate with two out of a dozen.'

The next job back at the homestead, after breakfast, is to check the pens filling every barn and alley. In a kind spring few lambs would be brought in from the fields, but this year a succession of them need reviving, or their mothers' milk supplementing. Proprietary lamb milk in dried powder form surpasses cow's milk for the job, though a goat is a useful lambing acquisition.

Yard chores finished, the shepherd sets off once more, perhaps taking two dogs this time, to drive out as many couples as possible. A good dog will 'set' a ewe not too keen on its lamb, arousing mothering instincts infuriatingly dormant until the watchful collie causes the sheep to stand and give succour to her offspring. The dog does not move, merely crouches or lies holding the sheep with its eye. After a few such sessions the lamb becomes strong enough to assert its rights.

Peak work in lambing time comes ten days to a fortnight after the first lambs are dropped. Births still come thick and fast, but there is an ever-growing number of young lambs to watch. At this stage, too, the first-born are ready for the fell. Before they leave the enclosed fields they must be 'marked up', with flock-mark, ear-tag and individual mark denoting their sire. The usual practice is to drive a score or two into a handy pen, mark, and set off with them to the moor gate, aided by two dogs. This is a satisfying job; the lambs have passed the first dangers of infancy, and valuable low ground is being cleared, but it takes time, and time is a very precious commodity at a hill lambing.

The afternoon round follows the same pattern as the second morning one, and then the day's last tour is planned to end as near dusk as possible. This does not always work in practice. A crop of mishaps may so delay the shepherd that he has insufficient daylight to check the last ewes properly. Then he relies even more than usual on his dog, finding perhaps a ewe in difficulties, when, with David Nixon; in 'Lambing Time' from *The Song of the Chilterns*

> Within gold arc of lamp
> my seeing fingers
> feel within the womb,
> to draw unwilling lambkin
> into light.
> And ewe's cry, raw
> with pain
> rattles the rafters
> of the witch black trees.

And stars which burn
a myriad holes
in cloth of sky,
look down and
trembling sigh.
As she new mother
turns her head,
to croon her love
and cosset with pink tongue,
her first-born into
leggy life.

Before the days of mechanical shears, communal clipping days were the rule. They still are, in some places. Neighbours go from one farm to another as the sheep are gathered ready for clipping and lend a hand until the task is done. No one reckons man hours. Clipping, however, inevitably clashes with haytime, and haytime is earlier than it used to be. School children in Bilsdale, North Yorkshire, always had an August haytime holiday towards the close of last century. Now most upland hay is cut in late June or July; it cannot well be earlier, as the meadows are filled with ewes and lambs until well into May, before growth has started on the hill.

Though clipping sheep by hand is hard work, the large numbers of helpers at a communal clipping transform it into a social occasion. There is a lot of chaff and back chat, and a great deal of gossip. A dozen or twenty people assemble, only about half of whom actually shear. Other essential jobs include catching the sheep, and dragging it to the shearer, for up-to-date handling pens are still lacking on many hill farms. The so-called catching pen holds a score or a hundred ewes awaiting their turn, and is filled periodically from the main flock by a combination of human and canine effort.

As the struggling sheep is hauled, hoofs scraping, towards the shearer, her bleats merely add to the constant hubbub of ewes separated from the lambs, lambs who do not realise that the disappearance of mum among more adults is only temporary.

73

Sheep smell mingles with wood smoke if horn burning (see Chapter 6) is done at this stage. As the fleece is finally clipped off, it is carried away by one of the wrappers, who flings it, skin side down, onto a slatted wrapping table. He flips its sides inwards, starts rolling from the tail, and finishes with a compact fleece tightly bound by its own neck wool pulled into a rope. Each fleece is either stacked on a trailer for later packing, or is pressed into a large canvas or hessian sheet supplied by the merchant. This sheet is suspended from either end, and takes a big weight of the harvest of the moors when finally stitched up and labelled.

The newly shorn sheep is dazzling white compared with the brown-grey of unclipped animals. She is not yet released. It is important to re-establish ownership, so on her way out she is given the appropriate colour marks. These must be positioned exactly. A blue stripe on the near shoulder becomes blue on the flank if the sheep jumps and the marker is careless. She may then be mistaken for a sheep of another flock. Branding is often the job of a retired shepherd, as it calls for exactness rather than strength.

Throughout the day, the wives, sisters, mothers and daughters of the men concerned play a vital part in sustaining spirits and energies. Meat pies, fresh bread and scones, tea and coffee are spread out while activity ceases for a few minutes, and new supplies of cold drinks are left for the sweating men.

As the sun lowers towards the circle of hills, sheep in the 'rough', their fleeces still attached, become fewer, while numbers of newly white become ever larger. The latter are reunited with their lambs in batches, and the cacaphony of sound rises to pandemonium. Each ewe calls for her lamb, each lamb cries for its dam, uncertain that the smooth-skinned creature answering its bleat is in fact the familiar and only provider of succour that it has known in its few months of life. Udders filled by hours of separation are drained, mother and offspring depart into a nearby paddock, but only after the watchful shepherds have ensured that every lamb is 'mothered up'. Careless or unskilled shepherding could easily result in a score of lambs which would never

Plate 10 Newly shorn North Country Cheviots at Cleuchhead, Bonchester Bridge, Roxburghshire. A large sheep, this is a breed of the grassy uplands rather than the heather moors. It has a higher lambing percentage than its Hill or South Country cousin, and ample milk for the lambs. Both types of Cheviot are mated to Border Leicester rams to produce the world-famous Scottish Halfbred. Tails of the 'Northie' are usually left three-quarter length, whereas the Halfbred is docked short because of the more lush pastures which are its home

recognise their dams again, and though they would now survive, they become 'light as corks' through subsisting on hill vegetation without milk.

On most hill farms these newly clipped ewes and lambs must be returned at least part way to their hefts that same evening. July grazing on the hill is seldom a problem. The congestion occurs on these small inbye fields hardly won from the surrounding moors, where lambing, tupping, and overnight keep for sheep needed the next day must be found. The more level fields are shut up as hay meadows from late May to July or August, so the flock-master is always conscious of the need to keep as few mouths near home as possible.

The trek starts. Fortunately, sheep drive best uphill and know where they are going, but the young lambs retard progress. Unused to dogs, they dash here and there, panic on losing their dams, and are only kept in the bleating mass by constant vigilance on the part of men and dogs who have been hard at work since early that morning, and will be so again as soon as the sun has completed its short dip behind the blue-black hills.

At last the moor gate is reached. The dogs form themselves into a loose semi-circle around the noisy flock now streaming through the gate from which every ewe takes its own path to the place it calls home. With it is its lamb, which if a female of sufficient quality will repeat the process in two years' time, its own lamb at foot. So it has been for hundreds of years, and so have tired men and dogs walked back to the lonely farmstead for a meal and a brief rest before the start of another clipping day.

WELSH MOUNTAIN RAM

8 RAM SALES AND TUP TIME

Though the ram sales are the climax of the shepherd's year, preparation for them is a continuing process. Selection starts by sorting the best ewes to match certain tups. In the breeder's mind at all times is a perfect specimen of his or her particular breed, and as no sheep is in fact perfect, matings are arranged to remedy slight faults. Ewes showing slackness behind the shoulder are put to a tup of outstanding conformation, those rather too light in face and leg colour mated to a tup somewhat on the dark side, while fleece improvements are sought by using tups of the correct fleece for the hill, which may be quite different from those best suited to a hill of different aspect and elevation.

These sires should be from a family line that has been consistently bred to a type. Their lambs are noted keenly at birth, for any young animal gives clues of its potential to the seeing eye at that stage, rather than weeks later when the effects of feeding have made their mark. All lambs pass through a stage when they look shabby, and have not developed the sought-after characteristics. In a well-run flock the best ram lambs off the best ewes get a distinguishing mark—a red pop on the tail head is one example—and are ear tagged according to their dam's number. As mentioned previously, her own pop mark indicates the ram with which she was mated.

In some breeds these tup lambs receive favoured treatment, but that should not be the object. Ram lambs are being reared to perpetuate a line under conditions limited to sparse herbage and wild, cold weather. It is no use pampering them and then expecting them to do a job on the hill, but at weaning they may go onto good pasture, as the next few months determine their bone structure. They are generally wintered with the help of a few concentrates, but return to the hill next May.

Shearing of these tup hoggs, as they have now become, is done early, to allow time for the coat to grow before the autumn sales, but June is soon enough for most. A tidy job is made, avoiding any unsightly tags, and then the sheep return to their high pastures. Throughout this time, numbers are being whittled down. A farm with eighteen or twenty rams to sell each autumn may begin with three times that number, and keep the best thirty at weaning before final selection.

A month before the sale, rams come down to better pasture, and for the last week are housed at night. This is more to prevent them from straying among the nearby ewes now coming into season than for the sake of the sale sheep, though they are better housed for a day or so immediately before the sale if the weather turns wet. A damp fleece spoils the appearance of even the best sheep. Another good reason for housing is that the rams are less disconcerted on entry into the ring, and stand smartly looking about them. If they have never been confined, they may panic and not look their best.

Some two weeks before the sale, each tup must be entered for the auctioneer's catalogue. Year of birth, sire's and grandsire's registered numbers are given, and whether bred by the owner. The sheep are protected against flies, for blow fly strike spoils the appearance, and legs are watched for sign of lameness. Tups have a habit of developing lameness just before their one season of use. On the day before the sale, each tup has a bit of dip splashed on, to add oil to the wool and lessen the chances of locks of wool being pulled up by the horns of other sheep in the pen. A 'peat dip', made by mixing peat and water, is a cheap and effective method of making all sheep in the pen appear the same colour. Legs and faces are washed with soap and water, though on some very dry, sound farms this is unnecessary. Other pre-sale jobs are to fill fuel tanks and load the vehicles with hay, straw and a tarpaulin if there is no covered accommodation; there will certainly be no time next morning. A rag to polish faces is included.

The great day dawns. Very early in the morning the shepherds rise, finish their essential stock chores and load their charges into the waiting transport. Tup sale morning has a ring of excitement about it that never palls. Among the oldest hands is a bit of that 'Sunday school treat' atmosphere, as their months of care are about to be tested against their fellows.

The journey to the sale in the half-light of an October dawn is marked by an unusual number of other vehicles, passing from the isolated farms whose dots of lights announce similar early morning activity. Through narrow, stone-walled lanes with sheep and cattle still resting on either hand, the transports converge on the market town where the auctioneer's staff is already astir. First job on arrival is to collect tickets from the office, each to be numbered according to catalogue and tied firmly onto a lock of wool on the sheep's back.

The show that takes place before the sale adds fire to the flames of competition. Any sheep gaining a prize ticket is assured of extra interest in the ring, but these pre-sale classes are a far cry from the sunny summer events and throngs in holiday mood. Here is a serious business, the lines of tups standing restlessly under stone wall and hawthorn bush, or in the cramped conditions of the sale ring if wet. Spectators are potential buyers; they are having a closer look at something that may influence their own flock for two or more sheep generations.

'Get your sheep pulled out', calls the steward, at which time Ray Dent admits that he frequently has not made up his mind which to show.

'We may have an outstanding sheep, which is no problem, but have not decided which of the next half dozen to bring forward. We haven't the man power to show them all, so there we are picking and choosing at the last minute'.

Ray is usually buying as well as selling, so must get round the alleyways in a hurry to look at possible stocks. He likes to see how an animal walks on its way to the ring; once there, it may skid about and proper observation of its action is impossible. A hill sheep is of no use if not of good action.

Meanwhile, son Ian is holding the fort and dealing with prospective customers. For this purpose he has kept a notebook of the pedigrees of all rams sold during the previous years, and has the breeding of all present sheep to hand. A buyer does not want stock closely related to those he already has, and Ian is able to advise which are related by consulting his notebook. Sometimes this loses a sale. Two brothers once entered the Glenwhelt pen, and went silently about their business of picking out two rams to match their ewes. They made their choice, only to be told that the two sheep were by the same sire as the ram they bought two years ago. They were interested in no others, so did not buy, but they came back the next year.

If Ray Dent finds a suitable sheep, he consults Ian and his employer. They discuss its suitability for a certain cut of ewes, and examine others in case they do not get their favourite. At a sale of hundreds of rams, only a handful may be outstanding, spotted by a hundred keen eyes. To find the ideal sheep that ninety-nine other shepherds have missed is as likely as turning up a nugget of gold when dry-stone walling on a Dumfriesshire hillside.

There are many things to do on sale day. The shepherd may remember a ram which took his eye two years ago—and here are his shearling offspring coming down the alley, feet pattering on the concrete, horns clashing and scraping. Are they as good as expected? Would one suit the home stock?

In adjoining pens the wiles of salesmen are apparent, for good salemanship has always played a big part in the armoury of successful livestock breeders. Each employs a different way of coaxing potential buyers into his pen, and vendors do like their wares to be examined. A show of interest in the sheep offered is a gesture of friendship, or at least an exercise in public relations.

Sales are in catalogue order, usually starting with a number drawn by ballot. A special stir of excitement accompanies the champion to the small pen where he awaits his entry into the ring. Experienced sales-goers take up vantage points well in advance, for though the hill shepherd leads a lonely life for much of the year, an astonishing number accommodate themselves in a small space when there is something exceptional awaiting the auctioneer's hammer. Entered alone onto the sawdust circle, the top sheep paws this way, runs another, lowers his head to charge, only to be diverted by a trio of ring stewards sufficiently long for the bidding to rise by hundreds, fifties, tens and fives. Blackface rams have topped £7,000, while three figures is a high price for a Herdwick, and a few hundred pounds for a Welsh Mountain ram.

As the last bids fall at the last tup sale, there is no desperate rush to get home. October darkness has fallen; a thin rain falls from clouds scudding across the moon, and there is stock to feed when the shepherds return to base. Yet, though they have mingled daily with their fellows during the past month, much remains to be argued and discussed. They know the long winter stretches ahead. What will it have brought before they meet again? Snow causing arduous work, raging blizzards burying the sheep so carefully husbanded over the seasons, or a pleasant, open time when the ewes forage for themselves with little recourse to the sweet-smelling hay won from summer meadows?

'I wish the tups sales went on all the year round', said one. But they don't, and it may be that the bustle and companionship have a sharper edge because of the long stints of isolation in between. For, shortly after the shepherd returns home, he will be so busy among his own sheep that social contacts will be limited to spare time. He will be preparing his ewes for mating and the start of another cycle. On the success of 'tup time' depends next spring's lamb crop. Ill weather then can still deplete it, but without a sound start and all ewes given a chance of the ram, no lambing can be a good one.

The ewe carries her lamb for five months, and lambing is timed to coincide with spring sunshine and grass growth. At least that is the idea, but British springs are fickle things, and there is no guarantee that the uplands will have warmed by mid-April. If lambing is delayed much later, however, the lambs have too little time in which to grow during the short period when hill herbage is at its height, and are too small to sell well at the autumn sales with their rather rigid dates. Some sort of compromise must be worked out, and early April on the lower hills, mid-month for the harder places and May Day for the really exposed farms are dates arrived at by generations of trial and error.

A hill ewe tupped on 5 November will lamb about 1 April. Therefore the end of November and early December is the time for loosing the tups, for starting once more the hill man's constant striving for survival and improvement of his flock.

On farms with enough lower hill or allotment ground, a proportion of the ewes are crossed with rams of other breeds. In the case of the Cheviot, Scottish Blackface and Welsh, these 'crossing' rams are usually Border Leicester, with Blue-faced Leicester, Teeswater or Wensleydale in the Dalesbred and Swaledale. Because of the slightly more sheltered conditions it may be possible to tup the 'crossing' ewes a week or ten days before the main flock, so that they are mostly lambed and out of the way. The disadvantage here is that this protracts the non-stop lambing routine, and working at such pressure with accompanying shortage of sleep is easier to bear for three than for five weeks.

Though the excitement of the sales is over, the days before tupping are exceptionally busy ones for the hill shepherd. He must go through his whole flock, and mark up the ewes to run with the different tups. The 'crossing' ewes destined to produce halfbred lambs are selected first. They include sheep below the best breed standard, possibly of wrong colour or carrying some defect of conformation that it is not desirable to perpetuate in the pure-bred flock. To these are added the youngest or oldest sheep; those between two- and four-shear are best able to with-stand the rigours of a hill winter, so the others receive favoured treatment. From the disease viewpoint, there is much to be said in favour of crossing the younger ages, as scourges like scrapie (see Glossary) do not appear till at least the second year, and a four-shear sheep has proved herself in a variety of ways.

Then the shepherd thinks of his best tups, which may include newly bought-in ones, and to them he draws suitable ewes, matching weaknesses in the ewes against the sires' strong points. This ability to carry sheep 'in the eye' is a mark of the good stockman. Each ewe receives a distinguishing colour mark accord-ing to her future mate, so that she may be quickly drafted into the correct batch and taken to a separate enclosure where she and her companions await their mate.

The next operation is to clip out the tails. The length of wool developed by selective breeding may make mating difficult, but this tidying round the tail root must be done in good time, otherwise the freshly-cut wool becomes bristly for a few days, and may damage rather than aid the tup.

After sorting the ewes, the various rams are walked to the different lots, or taken by van or tractor transport box if road conditions allow. On some hills each ram's chest is marked daily with a sticky rud or raddle (see Glossary) of bright red, blue or green. This smears the ewe's rump as he mounts her, and in-dicates to the shepherd that the ram is working all right. It shows potency, however, not fertility, so the shepherd must wait for seventeen days, or the length of the ewe's breeding cycle, to see if many ewes are being marked again by the new colour now used on the ram. If they are, that male is immediately taken up and another proven one substituted, for a prolonged lambing time is wearing on the man and upsetting to flock management.

Instead of the daily smearing with rud, a chalk-carrying harness may be affixed. This has the disadvantage that in wet, cold and windy weather, such as often accompanies hill tuppings, the straps chafe the ram's armpits and make him lame. They may also be said to encourage lazy shepherding as, with daily rudding of the rams, those magnificent animals must be caught and, possibly given a feed of corn. Again, other shepherds use no mark, but rely on observation that the ram is interested. He may not be seen working very often, for each lot of ewes is visited once a day only. They are gathered up round the flock sire if necessary. In an open autumn, they may be grazing far and wide, so this daily round up saves the rams' energies and gives each ewe the chance to mate.

The ram's chest colour is changed weekly if it is desired to know which ewes will lamb within certain dates. Where large numbers and a lot of walking are involved, however, there may simply be no time for such niceties, for late autumn days are short, and the shepherd must be stepping back downhill by mid-afternoon to re-enter the home yard by dusk.

Tupping time lacks the continued excitement, long hours and picturesqueness of lambing. Nonetheless it is the basis of a successful year. In each enclosure or section of hill, the first thing to check is that the correct tups are present, and that no breeding males from a neighbour's flock have strayed onto the heft. There is also the possibility that tups from another heft on the same farm have become mixed; the fact that a tup has the choice of seventy or ninety prime females does not guarantee that he will stay at home.

All this time the shepherd continues steadily on his long round, two dogs at heel, for each lot may have to be gathered to the tup. His worst enemy is the hill fog prevalent in this late autumn period, for to work sheep in it may do more harm than good. All he can do is wait for the weather to clear, and mend stone gaps while anxiously wondering if that new Wether Cote shearling bought last month is still with his ewes on the far fell. He also wonders if the offspring now in the womb will repay the money expended, but the answer to that awaits next spring.

THE ENEMY

9 HEATHER BURNING AND DRY STONE WALLING

The word 'shepherd' is defined in the *Oxford* dictionary as a 'man who tends sheep at pasture.' No British shepherd spends his time solely with sheep, however. A variety of other skills come into play throughout the year, among the most important of which is heather burning.

This is one instance where shepherd and keeper work together. Only the shoots of young heather make acceptable food for either sheep or grouse, and the only practical means yet devised of bringing about a continuing supply is by burning older growth in rotation. The rotation is from four to twenty years, according to the type of ground, for this invaluable plant varies considerably between different areas. A certain amount of older, rough heather is useful for nesting birds and to peep above the snow in winter, but, generally speaking, heather is not burnt as often as it should be.

85

One reason is that the operation may not be carried out when grouse are nesting. The latest date in most counties is 31 March, later in northern districts. 'The more sheep, the more grouse' is a well proved saying not always accepted by shooting tenants. It would be futile to pretend that shepherd and keeper always agree, but where they do not the fault usually lies with the employers. In particular, city syndicates of shooting tenants tend to believe that because they pay more rent for an area of ground than the farmer whose living depends on it, they may do as they please. Unfortunately they lack knowledge of the grouse except as a target from 12 August to 15 December, and try to reduce the number of sheep, the number of days allowed for shepherding and the amount of heather burnt. This attitude is the more tragic when it results in embittered relationships between two classes of professional countrymen who may be the only people living in sight of one another.

Heather burning should be looked on as a sport, said grassland pioneer Sir George Stapledon. And so it is, on a fine March morning, with a light breeze blowing, the heather tinder-dry, the air clean and crisp and two or three dales spread out below. Heather is best burnt in strips, so that sheep or grouse pass easily into the part desired and seek the various stages of regrowth. A birchwood beater and a box of matches are the only tools required. When burning near forest fence, it may be necessary to light a safety break first, for moor fires easily get out of hand. Badly planned afforestation has affected some hill farms adversely, as it is no longer possible to burn without risk of destroying the trees and being held responsible.

With the breeze at the men's backs a handful of dead bracken is set alight, transferred to the dry heather and from there carried along the line desired. Sizzling and crackling the flames advance, leaving behind a charred dark-blue mass of pungent-smelling ash. The smell of burning heather is the most nostalgic mnemonic for people from Scotland or the Isles far from home.

Workers beat the flanks, keeping the fire to its desired limits, or run back with snatches of smouldering flame to areas of ground

Plate 11 Heather burning on the Northumbrian moors. In the foreground, charred remains contrast with the bushy growth in front of the line of fire; next year will show new green shoots. Burning must be completed before the bird nesting season, and in a wet spring few chances occur. Rotational burning of heather helps both sheep and grouse, the other economically important species on the moor. Width of the blaze is controlled by beating out the flames on the flanks, as here

missed in the advance. Pillars of dark smoke arise, and may dot an entire upland county on a dry March day. If the fire moves too slowly, it sets alight the underground peat beds, which may burn for months. If too rapid, there is the risk of the flames getting out of hand. But when conditions are right, heather burning is one of the finest jobs on earth, for a great deal of good is being done by a primeval agency with the minimum of human aid.

It is not always like that. Many a spring is too wet for burning, so the rotation falls behind and even more must be attempted next year in an endeavour to catch up. One golden rule is never to leave a fire unattended, for there is no telling when the upland breezes will shift and burn all in front of them.

When shepherd and keeper are on good terms, the relationship is of the greatest mutual benefit. Both hate and fear the carrion crow, which attacks newly born or ailing lambs about the eyes and navel, older sheep fast in snow drift or gutter, as well as every grouse nest it can find. In the vast stretch of moorland beat, the shepherd may spot crow nests that the keeper does not, and informs him immediately.

While 'looking' the hill, the shepherd sometimes comes across a fox earth with a litter of cubs. A new one may be unknown to the keeper, and word passed quickly saves the lives of both nesting grouse and young lambs. The most usual method of dealing with foxes is for the keeper to put his terrier down the earth and wait above with his gun. In Lakeland, the six fell packs of foxhounds operate primarily to protect the lamb crop, each pack accounting for over 100 foxes yearly. Even so, the hounds merely preserve some sort of balance, for the odds are very much with the fox in the caverns and precipitous rocks of the Cumbrian mountains.

The shepherd and his family may well help the keeper by acting as beaters during the grouse-shooting season. In dead of winter, when storms blow really hard, the keeper saves many a sheep by helping transport hay during the short winter days. Both men work together at ditching or gripping, an operation designed to carry off surplus water and grow better vegetation for sheep and grouse. Today, most gripping is done by machine.

Dry-stone walling is a skill at which the hill shepherd must be adept. International and World Supreme Champion sheep-dog handler Raymond McPherson came to a Cumbrian farm from the north of Scotland, where walls are few. 'I soon learnt the art,' he said, 'I had to!' The thousands of miles of dry-stone walls on our hill farms are peculiar to Britain. Where the craft is practised overseas, it is usually in the former dominions or colonies where Britons took their skills with them. Dry-stone walls in France are mere piles collected from the fields and topped with wire; a Lancashire or Dumfriesshire stone wall is built to turn sheep for a century.

Frost, snow and hikers are the walls' worst enemies. Deep snow drifts piled against them and alternating frost and thaw

Plate 12 Dry-stone walls are essential as shelter and stock boundaries. Many hundreds of miles were erected during the seventeenth, eighteenth and nineteenth centuries, and keeping them in repair is a never-ending task for the hill shepherd. Here a new wall is being built above Otley, Yorkshire, on a farm famous for its Dalesbred and Teeswater sheep

inevitably cause gaps, as do careless walkers too lazy to find gates or stiles. The gap must be pulled right down to the base and rebuilt, using the biggest, squarest stones at the bottom. A skilled waller does not pick up a stone twice; he notes with his eye the exact shape needed, selects it from the pile and unerringly slots it into place.

Dry-stone dyking is the Scottish term for this art. It is by no means dying out; a thriving Register of Dykers has been compiled at Cally Estate Office, Gatehouse-of-Fleet, Kirkcudbrightshire, bringing together craftsmen and jobs, and notifying members of the many show classes now held up and down the country.

The walls snaking over our hills have many advantages. Besides making an effective stock barrier, they break the wind, providing the only shelter on many a windswept upland. They are fireproof, need no imported materials, and are virtually everlasting. Their snag is that they cannot be mechanised, so they take as many man hours per yard to build and repair as when they were erected perhaps two centuries ago.

It is remarkable how walls have been built up almost-sheer hillsides. Sometimes a huge boulder is cunningly incorporated in the line of the wall, sometimes the wall protects a sheep drop. 'Lunky' or 'smout' holes (see Glossary) are incorporated by using a long stone as lintel, and providing access for sheep only from field to moor. Having no mortar, the walls allow air to filter through and dry the coats of stock sheltering in their lee. A solid wall would be by no means as effective.

Walls stand on stony ground where no stake could be driven, and are founded on a solid base by removing the topsoil. Large flat stones slightly wider than the wall form the foundation, and on them the structure is built in two sides, the space between them being tightly packed with hearting—hard, small stones—as building progresses. Each course is kept level, the outside stones sloping very slightly inwards if anything. About two feet from the ground the first row of 'throughs' or 'thruffs' is built. These are flat stones which either fit the wall exactly or project slightly on either side, and hold the two cheeks together. One or more rows of throughs are used, depending on the height of the wall.

At the top the two sides blend into one and are surmounted by a row of coping stones, square or rounded flattish stones set on edge. They may be cemented on roadside walls, to prevent town thieves from taking them for rockeries. One square stone near a gatepost at Glenwhelt disappeared six times in one year!

Hill shepherd and keeper often combine also to keep open communications. Hill tracks suffer from heavy rain, floods and snow. No fleet of tarmac-carrying lorries repairs them. They are kept in passable order by the men who use them, with materials gained near the site. One Co Durham hill farmer digs gravel from the bed of a stream to repair his three miles of moorland road, using a tractor scoop for loading. His biggest difficulty is finding a suitable egress for his trailer. Potholes are made up by shovelling stones from the trailer floor, till the length is finished. Road repairs fall into the less urgent part of the upland calender, while tall stakes driven at intervals into the roadside mark its whereabouts when deep snow obliterates lesser landmarks.

Bridge building is another out-of-season task. Hill roads inevitably cross streams, and strong sleeper constructions are a vast improvement on the rather precarious plants which sometimes serve as crossing places. The shepherd must be skilled at such jobs, for his own safety and that of his flock depend on them when the weather is at its worst.

Plate 13 A shepherd's skills. A Weardale (Co Durham) sheep farmer welded these gates. Old hand tools and horse equipment augment the trap wheels. Surmounting the side gates are the skulls of two Swaledale rams, common in the area, while the small wheels on the posts are from horse ploughs. Hill dwellers have a whole range of skills outside their everyday work, yet often showing some connection with it

At one time not every shepherd's cottage had even a cart track to it. Today most are reached by a vehicle, even though the Land-Rover's four-wheel drive may be necessary. Groceries and fuel are delivered by the usual village tradesmen, although in some districts bought fuel is unnecessary; where deep peats lie to hand, the shepherd can provide his own winter supplies.

The usual time to dig peats is in late spring or early summer, with the hope of long sunny days to dry the slimy mass. In late May, lambing is virtually over, clipping and haymaking have not begun. A week among the peats in May is a 'perk' on some hill farms, and the farm tractor and trailer is brought into play to carry home the fuel. A cartload a week was once the tally where no other form of heating was supplied for the old-fashioned stoves, so considerable labour was involved. Now that oil and coal have become so expensive, peat cutting may again figure largely in the upland calendar.

The first job is to remove the top sods, and set them neatly on the previous years' cut alongside. The precision of this operation was traditionally regarded as a guide to a new shepherd's dedication and skill. The ground on the other side of the narrow strip to be cut is levelled and possibly burnt. Then cutting begins with a special peat knife, which has a cutting edge set at right angles to the main blade, so that flat bricks of peat are cut out in a single movement.

The men work in pairs. One cuts the blocks, the other is 'thrower-out'. He stands facing the cutter, and takes each liberated peat block in his two hands, flinging it expertly onto the bare turf above. The peats are 1–2ft long, 9–10in wide and 2–3in thick. They are laid out in a continuous carpet so that no weed grows between to hinder the drying process.

Peat is decayed vegetable material, cold, soggy, slippery and sour. To work with it in a biting wind is no joke, but on a fine day it is a pleasant enough activity. Through its water-absorbing capacities it is heavy when freshly cut, but loses weight and forms a 'skin' after a few days in the sun. When dry on the top, the peats are 'fitted' into long, inverted-V funnels, underside outwards.

Next stage is 'reckling'. Depending on the weather, this may be a fortnight or more after cutting, and is designed to combine maximum exposure to air with protection against rain. The bottom peats are set on edge, and others built on top in loose formation like a narrow hay cock or pike. The peats at this stage are so hard that they have formed sharp edges, and are only a fraction of their original weight.

Peat cutters have their own language. 'Ladies' peats' are wafer-thin, inadvertently cut by the unpractised digger. 'Saturday nighters' are solid, thick chunks suitable for the one time in the week when the shepherd is likely to have several hours' relaxation. Such peats keep the home fire burning with little refuelling.

The one disadvantage of a peat fire is the amount of dust it creates, but the joys are many. Well dried peats burn with a golden glow, give an even heat, and fill the room with a pleasant aroma. They awaken reminiscences of the sunny days when they were cut, however thickly the snowflakes whirl outside. They bring indoors memories of the cry of the peewit, the warbling curlew's call and the lonely piping of the golden plover as it flits from rock to mossy rock, ever in front of the walker on the hill. A peat fire is part of the hill itself; like the sheep and the people who tend them and whose livelihood they are, it evokes a feeling of being part of the scheme of things. No sheep talk betters that passed in front of an open peat fire, with a bevy of sheep dogs sprawled on the hearthrug, and outside the dark circle of hills where sheep safely graze.

10 THE SOCIAL CALENDAR

'If you work at a certain job, make friends outside it' is good basic philosophy. Yet the majority of a hill shepherd's outside interests have some bearing on his craft. With the passing of the autumn sales come dark evenings, and show committee meetings, breed society events and visits from friends and neighbours.

At one time visiting each others' homes was almost the only winter relaxation available to Dalesmen. This custom still retains its hold, and none are more sociable than hill shepherds after days spent alone. They enjoy simply being among their fellow men.

Evening classes are usually within reach, but painting, woodwork and metalwork are likely to have some shepherding connection. After a day on the hills among sheep, stone walls and grouse, the hill shepherd is likely to paint his favourite breed in a stone-walled enclosure, with a covey of grouse in the background.

The craft of crookmaking or stick dressing began in shepherds' cottages. Now it has spread to all sorts and conditions of men. David Grant wrote in *Shepherds' Crooks and Walking Sticks*:

> Shepherds' cottages of those [pre-1914] days were primitive affairs, often with nothing better than a paraffin lamp for lighting. This was barely sufficient for reading, yet adequate to see to carve a crook. It had the further advantage that heat from it could be used to bend the horn. In those days before wireless, television, or a daily paper in outlying parts, mid-winter evenings must have seemed long indeed. The hill shepherd whiled away the time by whittling at his crook.

The crook's shank is of wood, usually hazel, ash or holly, but so much land has been block-planted with conifers by the Forestry Commission and investment companies that it is by no means as easy to find as good a supply of shanks as formerly. The head is of wood or horn. Shortage of good horn is a strict limiting factor in stickmaking, and here the shepherd is uniquely placed. He may be able to preserve for all time part of a favourite ram by securing the horns when the old stalwart's days on the hill are done, and carving them into a crook to help cover the miles and catch that ram's descendants.

Shanks are cut at least one year before intended use, and hung up to dry. The first task then is to straighten the shank, by heating any bends. Tin foil wrapped around the stick prevents damage to the bark, and gentle heat from a lamp is best. The shank must be absolutely straight, or a first-class stick will never result.

When making a wood stick, the next process is to take off the sides of the head. Two flat surfaces are left, on one of which is drawn the head in outline. The distance from nose to heel of a crook is 3½–4in—less for a walking stick. The inside is always drawn first, then the outside follows naturally. The inside mark is now cut out with a bow saw. A Surform file or pocket knife is used to round off the edges to fit the hand, after which the line of the top is shaped. As much of the delicately shaded natural bark as possible is left.

Plate 14 Stages in the shaping of wooden shepherd's crooks

For a horn stick or crook, the shank is fashioned in the same way, except that it is left straight to join with the horn head. On the ram the horn is bulky and rough, far removed from the light, artistic head it is destined to become. The first stage is to saw the horn base squarely, giving a right angle with the proposed heel. The back of the heel is then set in a vice and filed as straight up and down as possible, rough horn from the inside being removed.

At this stage the horn is still as curly as when it came from the sheep. It must be turned and flattened, which can only be done under heat. An oil lamp is most often used, and the horn is heated evenly in the desired part, starting at the heel and working towards the nose. A vice gives the necessary hold, while the bends are taken out by using a long metal cylinder or a large pair of grips on the nose.

The reshaped horn is tied securely in place while still in the

vice, and left to cool. Only then is further straightening decided
on, but the nose must be absolutely perpendicular and in line
with the heel. The next stage is to shape the inside. A coping saw
followed by the cylindrical Surform file is brought into play.
The good stickmaker tries to keep clear of the 'white', or softer,
immature part of the horn.

Shanking, or fitting head to stick, is the next and vital part. The
top of the shank is shaped into a cylindrical peg, and a hole of
identical size bored into the head. The joint may be strengthened
by inserting a steel peg. A ferrule to join head and shank is
fitted into place, stag, sheep and cow horn all being both strong
and artistic, while brass, copper, aluminium and even alkathene
piping are used for a working crook. The shank tip also receives
a brass or rubber tip, to prevent wear.

Now come the final processes, when the natural shades of
wood and horn are brought out with an artist's skill. Instead of
saw and file, sandpaper, glasspaper and soft cloths are used.
Sandpapering until the horn is absolutely smooth may take
hours, but the long-lasting result makes it worthwhile. Poly-
urethane varnish lends gloss and protection to the finished article.

The owner's name may be carved on the head by leaving horn
or wood for the letters, and carving out the space around them.
This background is then painted red or black. Fancy sticks with
fish, otter, sheep dog or rabbit are additions to the popular
Scotch thistle at the nose, but all have one feature in common;
they are carved from the bulk of the head, be it wood or horn,
never stuck on afterwards.

While the shepherd is carving a stick highly polished enough
to enter at next summer's shows, his wife and daughters take
part in a variety of Women's Institute activities. If snow falls,
hill folk may have to walk, as did their ancestors at all seasons.
But snow four times the depth needed to cancel suburban
activities seldom brings Dales social life to a halt. 'In the spring
of the year we try to widen our field of visits before lambing',
said a Dales shepherd. 'Lambing seldom begins before late April,
so as nights become lighter, journeys become longer, though

there will still be a domino or darts drive in the local pub with half a pig as first prize!'

Lambing takes the whole of the shepherd's thoughts and energies for three weeks, ending about mid-May, and then comes a most interesting time of the year. Sheep-dog trials are held on light evenings or Saturdays, and the spring shows get under way. These are still comparatively recent and local, forming a grand get-together after the restrictions of winter and arduous lambing routine. They may not be good for the sheep, which are brought into show condition at an unnatural time of year, but they are great fun.

Spring shows to some extent replace the Shepherds' Meets. The secondary objective then was a 'crack' (chat) and a dram,

Plate 15 The 'Swaledale Royal' as the annual show at Tan Hill is termed. Claimed England's highest pub, at 1,732ft, Tan Hill is a white landmark over thousands of acres of open moorland near the Durham, Yorkshire and Cumbrian boundaries. Only Swaledale sheep are shown, and weather is invariably bitterly cold for this last Thursday in May fixture. Here a class of rams is being judged. Despite its isolation, a large crowd assembles at an inn that once welcomed drovers on their trans-Pennine journeys

with domino-playing long after official closing time. Then the shepherd and his dogs had to return home with the sheep he had retrieved. Journeying over the fell tops in darkness with thirty or forty sheep after an evening's jollifications is no light task. Cumbrian tup hirings are rather similar, the sheep being let for the season, then returned to their owners.

After one of these meets a party of Lakeland shepherds set off, in the days before motor cars, to walk a number of tups back across the fells above Buttermere. A thick mist covered the tops and, at a parting of the ways, even these hill men were undecided on the correct route. They set off in one direction, scarcely able to see a woolly Herdwick head five yards away. Their charges took that track most unwillingly, though pressed by men and dogs, until after some time it was apparent that the party was lost. The men could do nothing but give the tups their head, and these old stagers returned to the cross roads, aimed without hesitation in a different direction altogether, and arrived back above their own farmyard.

In summer the round of shows proper begins, and on most Saturdays there is one within range. The shepherd may be showing or judging, or simply interested in the cattle and ponies, produce and crafts of the locality. Such shows have a flavour of their own, and are a 'must' for any visitor wishing to sample the true character of upland communities.

After the shows come the autumn sales, starting with the halfbred lamb sales. Then the ewe sales follow, and finally the tup sales, when hill sheep farmers 'go on their holidays' as one put it, and attend sales of their breed almost daily for two or three weeks.

On a September Sunday before the Hawick ram sales in Scotland's oldest market, an event takes place which is a typical example of the sheep man's propensity for mixing business with pleasure. 'Tup Sunday' was initiated early in the century—Geoff Robson of Hexham has only missed one in fifty years—as a means of showing sheep away from the bustle of the sale ring. Prospective buyers visit each other's flocks, starting at Hindhope and

Plate 16 Preparing Masham ewe lambs for sale

Upper Hindhope deep in the Cheviot hills. There is a get-together
of breeders who may not have seen each other for a twelve
month, Border hospitality at its best, and a tour of old stone
farmhouses with outer walls hardened by decades of storm,
inner ones bedecked with hunting prints, warmth and good
cheer everywhere. After a suitable interval the cry of 'Tups!'
summons the company into the breezy sheep yards to inspect
the two-shear rams, the usual age for selling Hill Cheviots.

Next call is down the Kail valley to Chatto, where shepherd
John Douglas contrives massed colours of dahlias in his front
garden. Lunch at a Yetholm hostelry is followed by further

drives up winding valleys to Cliftoncote and Attonburn, mentioned in the early flock books of the 1890s. Afternoon tea at Mowhaugh—which surpasses luncheon at most London hotels—fortifies the party for another long drive to Kilham and finally to Elsdon Burn in the College Valley. This is the country of the old Border raiders, and as the moon rises over those valleys and the wind moans through the trees it is easy to imagine the dark shapes of horsemen looming over the hilltops. Names change little in the Cheviot world; Douglas, Elliot, Robson, and Scott

Plate 17 Tup Sunday in the Cheviots. Two-shear (two-year-old) rams of the Hill or South Country Cheviot breed being put through their paces at Hindhope, Jedburgh. Note the quality of the stone walls, and the vast expanse of grazing in the background. Hawick ram sales take place on the following Tuesday, and here potential buyers travel from one flock to the next to assess sheep on their home ground. A good, bold eye is sought in a hill ram, indicating character to lead his flock to safety when winter blizzards strike the uplands

figure in early sale catalogues just as today. Tup Sunday invariably ends around the huge table at Hethpool, with a vast meal and sheep talk continuing till the early hours. Tup Sunday to the band of Cheviot breeders compares with the Lords Test to cricketers.

WELSH BLACK EWE

DM>EM.ALDERSON
1977

11 SHEPHERDING AS A CAREER

'Don't look only on the glamorous side' is the first stipulation
for any prospective shepherd or shepherdess. Idyllic pastoral
scenes of tending lambs on sunny slopes are only part of the story.
Half a shepherd's life is spent in dirty, and often wet, clothes.
Hours are longer than in most industries, and pay lower. There
is no overtime for the hill shepherd; it would be quite impractical
to work out a system fair to either the employee or his boss. Yet
for periods of the year very long hours are worked, and there is
no dodging them.

If the prospective shepherd can stand that outlook, and not be
put off by bad weather, he should experience the dirty side of
shepherding before embarking on a career among sheep. He
must appreciate that sheep have to be cleaned under the tail,
that to treat their feet against foot rot is a smelly process, and
that they may be infested with maggots. Cleaning a 'struck' sheep
is a fair test for the potential stock man.

Weather on the hill is likely to be colder and wetter than in the vales. The weight of rainfall in areas like Snowdonia and the Western Highlands becomes crushingly depressing to man and sheep alike. Trudging about in oilskins day after day is a wearisome process, and as yet the shepherd must provide his own hard-weather equipment. Boots and leggings become ever more expensive, and have a comparatively short life on the hill.

Shearing is such a dirty job that the oldest of old clothes are saved throughout the year, worn for the requisite number of days and then burnt. Dipping is not much better; few dippers are splash-proof, and the liquid is detrimental to clothing. Working in plastic weatherproofs, the shepherd sweats and is unable to dry out, increasing the risks of rheumatism in later life.

Yet all these disadvantages are forgotten when the sun shines. 'On top of the hill, with the grouse becking, curlew calling and all the dale at my feet, I wouldn't change places with any man', said one shepherd. Though often working alone, the hill shepherd is seldom lonely. Men of the hills can see what their neighbours are doing in a way denied the lowland farmer. A tractor stuck on the opposite side of the valley is a subject for sympathy, possibly for practical help. Each farm, field and allotment within sight is known, as are the people working it, their dogs, and the quality of their sheep. Tom Jones is moving one lot of ewes and lambs into Green Pasture, James MacKinnon and his son Alec are mending a gap in the wall between Lark Field and the open hill, while the Scotts are driving home their Galloway cows and calves for subsidy inspection.

Even on a wet day of poor visibility, the shepherd has his dogs. They, more than ever before, are one of the great attractions of the life. The Border collies have become live to audiences of millions through *One Man and His Dog* and the 'Shepherd of the Year Award' on TV. 'Today twenty cars will stop to watch our dogs at work, for every one a few years ago', says Ian Dent. The visitors see the shepherd leaning on his crook on the hillside, possibly with pipe in mouth, effortlessly guiding the efforts of two collies in command of a swirling mass of sheep. They wax ecstatic over woolly lambs, without realising the hours of patient training that make up the first spectacle, or the fact that a proportion of the lambs are born for slaughter.

If, despite all this, a shepherding career is sought, how does the outsider set about it? The first thing to realise is that in much of Britain there are few shepherding jobs pure and simple. Even in Scotland the shepherds are required to take on more and more cattle, so that prospects of a 'sheep only' job are remote. Besides, it is better to command a variety of skills. The farmer may die or the farm change hands, and the shepherd is then left without a post through no fault of his own.

The beginner's best approach is to obtain a job as a general worker on a farm where sheep are kept. If he can arrange to help with the sheep at busy times, so much the better. If not, he must observe, talk sheep, and work among them in his own time.

The youngster must make up his mind what type of shepherding he wants. Hill shepherding is as far apart from lowland and, especially, intensive shepherding under cover as is dairying from egg production. Marginal-land shepherding lies somewhere in between, but it is of no use for a youngster with ambitions of a job on the Cheviots or Cairngorms to train on Romney Marsh.

Ideal training is as shepherd–stockman, taking in tractor driving and proficiency with machines associated with stock, such as hay balers and tedders. Then, and not till then, is the right time to embark on a course at a farm college with a specialist sheep course. One of these is at the Northumberland College of Agriculture, Kirkley Hall, Ponteland, where Principal Philip Blake stresses that 'There are no short cuts,' the shepherd must learn everything he possibly can before, during and after his one-year course.

The Sheep Management Course at Kirkley Hall is the only full-time residential course on sheep and beef husbandry in the UK. It lasts from September to the following July, totalling thirty-eight weeks of study time. A certain amount of practical experience, ability and enthusiasm are the factors sought from candidates. Of their thirty-eight weeks, students spend three on a lowland lambing and a further, later, three weeks on a hill lambing. Of the whole course 60 per cent is of a practical nature. Systems of management for the production of breeding stock, store and fat lambs are studied, as are sheep-farming economics and the fitting of a sheep flock into the rest of the farm. Wool production and shearing, and feed requirements of the various classes of sheep receive attention.

Sheep-breeding studies include a look at the new breeds, while disease prevention and control and the latest ideas on reducing labour on the stock farm are combined with dry-stone walling, hedging and fencing under expert tuition. By way of contrast, students learn gas and electric welding. The college flock of 400 ewes is augmented by co-operating farms in the north of England where students gain further practical experience.

The course leads to the College Certificate in Sheep Management, and members are entered for the Advanced National Certificate in Agriculture. Previous attendance at agricultural courses on a full- or part-time basis is a help. On completion, the shepherd or shepherdess should be competent to manage a large sheep unit, and there is a growing demand in the sheep industry for suitably trained people well versed in modern ways of flock management. The farmer's son returning home to manage the family flock gains equally from the experience.

'The present trend in farming is towards concentration and specialisation', says Philip Blake. Fewer enterprises on larger holdings have led to a need for just the type of shepherd which Kirkley Hall plans to turn out. That the practical side is adequately covered is shown by the kennelling of trainees' dogs on the premises, the principles of dog training being a very important part of the course.

Cost of tuition, board and lodging total £668.60, and most local education authorities make grants which will allow any suitable applicant to attend the course. In 1976, seventeen students were enrolled, and to date there have been sufficient jobs as shepherds to flocks of 4–600 lowland ewes and 400–1,000 hill ewes. Assistant shepherds' posts on still bigger units have also been obtained.

Whilst still at school the really keen youngster should take every holiday opportunity to watch what is happening among sheep. He can read about the sheep's anatomy—how its stomach works, its breeding cycle, its wool growth. He can study soils, fertilisers, minerals and vitamins, all things having direct bearing on his future charges.

While shepherds' sons start with a tremendous advantage, and are half trained by the time they leave school, sheep specialists cannot be recruited from this source alone. There is a time lag, for few entered the sheep industry during the depressed 1960s. More trained people are needed to meet potential expansion, and shortage of them puts a brake on that expansion that can play so vital a part in Britain's future.

'As soon as possible, youngsters need a certain amount of responsibility', said Ray Dent. 'When I was fourteen, my father gave me 170 ewes to lamb in one allotment. "If you get stuck, give Joss Ralph a whistle over the wall", I was told, and I was very proud indeed to get through that lambing with little loss.'

A lambing shepherd washes his hands and scrubs his nails a score of times daily. If that doesn't appeal—don't become a shepherd. If regular hours are sought, keep well away from sheep. Ian Dent had finished his routine tasks on Swinhope Fell as dusk was falling on a foggy winter's afternoon. He was looking forward to his tea and his evening with his friends. Walking down-hill past a disused house and still a mile from home, he heard a sheep 'blaring'. Setting off across country, he traced the call to a sheep that had become blind—a not uncommon winter affliction. If left till morning, that sheep might be dead. She couldn't be driven by dogs she couldn't see. She could hardly walk. For half an uphill mile she had to be alternately dragged and carried till Ian's father arrived to check on his son's absence. A Blackface, Welsh or Swaledale ewe may not weigh a hundredweight on the scales, but wet fleece and dragging limbs add to the burden with every step.

Every responsible employer understands such situations, and knows that all is well if his shepherd is apparently not doing very much. 'One week in July we may be shearing late every evening, and when the job is finished I take a day off if I feel like one', said Ray Dent. 'I might go to Hexham market, on the day that my boss happened to be there. He would never dream of saying: "What are you doing here?" He would say: "Come along for a meal, or a drink", and ask how things are. Shepherding would be a completely impossible profession if the employer was never off one's back, and did not allow the shepherd a certain amount of responsibility and decision-making, no matter how young he or she is.'

It is as important for the employer to understand the rigours and uncertainties of looking after sheep as for the shepherd to understand hill-farming economics. Team work of a type rare in other industries is essential.

In Scotland it is common for each shepherd to own so many 'pack sheep'. These are his own prerogative, and graze with the rest. Though limited in numbers, there is no restriction on quality, and sale of a few tups, perhaps running into three figures, is a substantial boost to income.

Management promotion is another step up the farming ladder, but the number of managerial posts open to the sheep specialist as such is likely to be limited. The whole question of farming on one's own involves capital and availability of farms. A move towards smaller units may become a national necessity, but is something that cannot be prophesied.

DM·EM ALDERSON
77

GLOSSARY OF SHEEP TERMS

Allotment Walled enclosure, usually on a hillside. The ground between the enclosed fields and the open hill was 'allotted' amongst the various owners of grazing rights.

Barren, geld, eild A ewe not in lamb when she should be.

Broken-mouthed Older sheep with loose or missing teeth, unable to thrive on hill pastures.

Cast And old sheep fit only for the butcher. In some areas a sheep fast on its back.

Couped, rig-welted A sheep turned onto its back and unable to rise.

Crone Old ewe.

Crop Preceded by a number this indicated the number of lamb crops borne by a ewe.

Dagging Clipping soiled wool from around the tail.

Draft ewe At four, five or six years old the ewe is 'drafted' from the hill for further breeding on the lowlands, and becomes a draft ewe.

Ewe Adult female. Often linked with age definition, eg ewe lamb, two-year ewe.

Fell, hill, moor Regional names for grazings of natural herbage, possibly unfenced.

Gimmer Young female (gimmer lamb is a ewe lamb). Gimmer shearling is a young ewe between its first and second shearings, though the Borderers simply refer to them as gimmers or **dinmonts**).

Heft An area of the hill grazed by certain individual sheep. No precise number, but often one to two hundred.

Hefting, hefted The verb for accustoming sheep to stay on their own territory. Originally very hard work done by constant shepherding and active dogs. Now a habit acquired by the ewe lamb from her dam, reinforced by the shepherd on his rounds.

Hirsel A number of hefts, comprising the area of ground herded by one shepherd, and the sheep on them.

Hogg, teg (also **hogget, hoggerel**) From December of the sheep's first year to shearing. Hogg more common in north, teg in south.

Inbye The fenced or enclosed lower fields of a hill farm.

Lamb Sheep until weaning or the end of the year in which it is born.

Lunky hole, smout hole Low hole in a stone wall, big enough to take one sheep at a time, linking field and hill. Easily blocked when not in use.

Polled Hornless.

Ram, tup Interchangeable terms for an uncastrated male. Often linked with age definition, eg tup lamb, shearing ram.

Rud (or **raddle**) Coloured powder to which liquid is added to form marking paste.

Spain, wean The act of taking young stock from their dams. End of the milk-feeding period.

Strike, struck Shepherds' term for fly attacks, the eggs hatching into maggots which literally feed on the animal.

Theave as gimmer.

Twinter A Lakeland term for sheep in their second winter. On the hardest Lakeland fells, ewes do not produce their first lambs until they are three years old, compared with two years old on most hills.

Two-shear Sheep in its third year. Hill sheep are shorn first at just over one year old.

Two-tooth as gimmer.

Wether Castrated male, for fattening. Wether lamb, wether hogg, shearling wether. The single word 'wether' is more usually associated with older sheep kept to four or five years old for their fleeces. They marked the hill boundaries, and were useful leaders in a storm. Modern economics are against them, but they are still kept on land in especially difficult areas. Some outlying hirsels were 'wether hirsels', too remote and high to be suitable for a breeding stock.

SHEEP DOGS

Doubles, brace Running two dogs simultaneously.
Drive Dog sending sheep away from the shepherd.
Eye The power to control a sheep without touching it. The natural instinct to work.
Fetch Dog bringing sheep towards the shepherd.
Gather Collecting sheep from scattered grazing positions into a compact bunch.
Lift The brief period when a dog has stopped at the end of its outrun, before bringing the sheep forward. Important as it gives sheep their first impression of the dog.
Outrun Course taken by dog in gathering the sheep.
Pen Small enclosure with or without a gate, into which sheep are driven.
Shed, shedding Parting certain specific sheep from the others without the use of holding pens.
Single Separating one sheep from the rest.
Strong dog No reference to physical characteristics. Determined, courageous, master of difficult sheep.
Weak dog Again no reference to physical attributes. Liable to give way to awkward sheep, lacking power to work a large flock.

SCOTTISH TERMS

Bothy A cottage or single room away from the farmhouse.
Bucht, fank, ree A sheep or cattle fold, a sheep pen.
Corrie A hollow in hill or mountain side, where deer often lie.
Pad, rake A path or beaten track. To rake sheep is to move them downhill each morning and uphill each evening.
Scree Debris or loose stones on a mountainside.
Shiel, sheelin A hut or cottage, particularly for use during summer grazing distant from the homestead.
Stell A circular pen out on the hill, where sheep gather in a storm. Driven snow funnels round a circular pen, leaving the inside clear.
Strath A valley or plain through which a river runs.

APPENDICES

A: SOME COMMON SHEEP DISEASES

Black disease A rapidly fatal affection of adult sheep, especially those in thriving condition. Associated with liver fluke.

Braxy Commonly attacks hoggs, especially in autumn and after a hoar frost. The sheep are usually found dead. Vaccination controls this disease.

Foot rot A contagious disease caused by a specific microbe entering the soft, sensitive foot tissues and causing lameness. Injections, foot baths and sprays are all used against it. There is no foot rot in Iceland, despite muddy areas with which the disease is commonly associated.

Head fly A distressing affliction that has become more prevalent in recent years. These flies constantly buzz round the head and give the sheep no peace. Various dips and creams are being tried, but the ban on dieldrin has worsened the position.

Joint-ill An infection in lambs occurring at, or shortly after, birth. Dullness, loss of appetite and swollen joints are symptoms, cured by drug treatment if caught early.

Lamb dysentry Highly contagious. Affects very young lambs, but now controlled by vaccination.

Liver fluke Causes much loss. Parasitic worms or 'flukes' invade the liver, and cause varying degrees of weakness. Primary control is to get rid of the host snail of the parasite. Veterinary control measures are available.

Louping ill or **'trembling'** Recognised for almost 200 years as a cause of serious loss. Caused by a virus transmitted from sheep to sheep via ticks to which some upland areas are subject. Vaccine prevention combined with regular dipping controls the condition.

Pine A wasting disease caused by lack of cobalt in the pasture. Mineral licks and powders containing cobalt are one control, dosing sheep by mouth another.

Pulpy kidney Affects lambs and sometimes older stock, usually in thriving condition. Caused by a soil-borne microbe. Controlled by vaccines.

Scab Caused by mites which bring about intense irritation. A notifiable disease, sheep scab was with us for a very long time until cleared after World War II, and has recently come into prominence through its reintroduction from Ireland. Three hundred years ago the Robsons of North Tyne raided Liddlesdale and took sheep which proved to be infected with scab. They returned, hanged seven Grahams and added: 'Gentlemen cam to tak sheep, they were not to be scabbit.'

Scrapie A disease of the nervous system with no known cure and complicated by a long incubation period of two or perhaps many more years. Breeding programmes are now bringing control.

Swayback Caused by a shortage of copper in the bloodstream. Affected lambs cannot walk or co-ordinate their limbs properly. Correct feeding of in-lamb ewes prevents the disease.

B: USEFUL SHEEP-FARMING ADDRESSES

Animal Breeding Research Organisation, West Mains Road, Edinburgh
British Wool Marketing Board, Kew Bridge Road, Brentford, Middlesex
Country Landowners' Association, 16 Belgrave Square, London SW1
Grassland Research Institute, Hurley, Maidenhead, Berkshire
Hill Farming Research Organisation, Bush Estate, Penicuik, Midlothian
Institute of Animal Physiology, Babraham, Cambridge
International Sheep Dog Society, 64 St Loyes Street, Bedford
International Wool Secretariat, Wool House, Carlton Gardens, London SW1

Joint Sheep Group, National Farmers' Union, Agriculture House, Knightsbridge, London SW1
Meat and Livestock Commission, Queensway House, Bletchley, Milton Keynes
National Institute of Agricultural Botany, Huntingdon Road, Cambridge
National Sheep Association, Jenkins Lane, St Leonards, Tring, Herts
National Union of Agricultural and Allied Workers, 308 Gray's Inn Road, London WC1

Breeders' Associations

Blackface, 24 Beresford Terrace, Ayr
Black Welsh Mountain, Summer Lane, Combe Down, Bath
Blue-faced Leicester, Blencogo, Wigton, Cumbria
Border Leicester, 24 Beresford Terrace, Ayr
Cheviot, 5 Tower Knowe, Hawick, Roxburghshire
Clun Forest, 11 Blackfriars' Street, Hereford
Colbred, Crickley Barrow, North Leach, Cheltenham, Glos
Dalesbred, 45 Allhallows Lane, Kendal, Cumbria
Dartmoor, Bilberry Hill, Buckfastleigh, Devon
Derbyshire Gritstone, Red Lees Road, Cliviger, Burnley, Lancs
Eppynt Hill and Beulah Speckled Face, Market Street, Builth Wells, Powys
Exmoor Horn, The Avenue, Minehead, Somerset
Herdwick, Glenholm, Penrith Road, Keswick, Cumbria
Hill Radnor, Newmarket Chambers, Abergavenny, Gwent
Jacob, Jenkins Lane, St Leonards, Tring, Herts
Kerry Hill, Milford Road, Newtown, Powys
Llanwenog, Bertheos, Creuddyn Bridge, Lampeter, Dyfed
Lonk, Jack Hey Lane Farm, Cliviger, Burnley, Lancs
North Country Cheviot, Bellfield Road, Kessock, Inverness
Rough Fell, Milnthorpe Road, Kendal, Cumbria
Shetland, Fairview, Vidlin, Shetland
South Wales Mountain, The Bryn, Pontllanfraith, Gwent
Swaledale, Spring End, Low Row, Richmond, North Yorkshire
Teeswater, Edengate, Warcop, Appleby, Cumbria
Welsh Mountain—hill, c/o WAOS, Brynawel, PO Box 8, Aberystwyth, Dyfed
Welsh Mountain—pedigree, Tudor Cottage, Llwyndafydd, Llandysul, Dyfed
Wensleydale Longwool, 27 Fountain Street, Ulverston, Cumbria

White Face Dartmoor, Sawdye and Harris, Newton Abbot, Devon
Welsh Halfbred, Brynawel, Aberystwyth, Dyfed

Manx Loghtan,	
North Ronaldsay or Orkney,	combined flock book, Lower
Portland,	Eastrip Farm, Colerne,
St Kilda,	Chippenham, Wilts
Soay,	
Whiteface Woodland	

C: THE GEORGE HEDLEY AWARD

The George Hedley Award was instituted in 1960 by the National Sheep Association. It honours the memory of a prominent Roxburghshire and Peeblesshire sheep farmer named George Hedley who was Association Council Chairman at the time of his tragic death in a road accident on his way to attend a London council meeting. A suitably inscribed medallion is presented annually to a person who has given outstanding service to the sheep industry. Presentation is at NSA's Three Day Event in early summer.

Recipients to date are:

1961 Dr R. F. Montgomerie, director of the Veterinary Research, Wellcome Foundation Ltd, for work leading to improved methods of control of liver fluke, nematodirus, pulpy kidney and other diseases.

1962 A. R. Wannop OBE, first director of the Hill Farming Research Organisation. He was largely responsible for building up a soundly based research body specialising in hill farming and the use of marginal land.

1963 Oscar Colburn, farmer, for pioneer work in sheep breeding.

1964 Dr J. T. Stamp, director of the Moredun Institute, Edinburgh, for research on sheep diseases, particularly scrapie, enzootic abortion and Johne's disease.

1965 Dr Walter Downing, for research on control of external parasites and dipping techniques.

1966 H. B. Parry who researched scrapie and developed recorded health systems.

1967 R. T. Rowlands OBE, who served the interests of Welsh sheep farmers.

1968 Dr Allan Fraser, Rowett Research Institute, Aberdeen, and University of Aberdeen.

1969 John Drinkall CBE, chairman of the British Wool Marketing Board.

1970 John Chilman MBE, Clun forest sheep breeder and reclaimer of marginal land.

1971 Prof Colin Spedding of the Grassland Research Institute who researched grazing procedures and the sheep's nutritional requirements.

1972 P. Stewart Tory, leading breeder and exporter of Hampshire Downs. Overseas judge.

1973 Dr J. M. M. Cunningham, director of the Hill Farming Research Organisation who, by the written and spoken word, has reached a large circle of sheep men.

1974 Gwynn L. Williams, closely involved in recording, progeny testing and performance testing of Welsh Mountain sheep.

1975 Douglas McDougal for extensive service to the sheep industry.

1976 Joe Raine, breeder of Blue-faced Leicester and Swaledale sheep, active on many national sheep organisations.

D: SHEPHERD OF THE YEAR AWARD WINNERS

The first winner of the Shepherd of the Year Award was Miss Dorothy Bell, Meadow Flatts, Alston, Cumbria. There she herded 1,800 Swaledale ewes and 500 ewe hoggs, and was the only assistant to her employer Mr Tommy Elliott.

Meadow Flatts is itself a high farm, but the real work began at Priorsdale six miles away, where the holiday cottage homestead stands at 1,740ft. Much of the 4,000 acres of fell tops 2,000ft, the highest point being almost 2,500ft above sea level.

On leaving school, ten years before gaining the award, Dorothy Bell had a real baptism of fire during the 1963 winter. The fell road was completely blocked for twelve weeks, and on every single day this shepherdess walked twenty miles to bring the sheep onto the tops, and stayed with them while they grazed herbage from which the snow had been blown clear.

Her nominator, Douglas Dickinson, said: 'During last winter she was out at 2 am (when a sudden snow storm blew up during the night) and consequently she lost very few sheep because of storm, on a farm with a history of high snow losses. She has been able to obtain returns comparable to farms 1,000ft lower.'

Chairman of the judges Professor Gordon Dickson spoke of Dorothy Bell's judgement, expertise and dedication, substantiated by her success in the 'cut-and-thrust' world of the northern show yard and sale ring. 'The rigours of her work, however, have in no way detracted from her charm and vital personality. She is an inspiration to young people contemplating a career in shepherding.'

By contrast, runner-up Mr Tom Ellis Roberts is a shepherd in the classical mould, who has devoted a lifetime to shepherding. In addition to his 800 ewes and 100 ewe lambs at Llysfasi College of Agriculture, Ruthin, Denbighshire, North Wales, his working days are filled with bringing on a succession of young people interested in sheep. Complete master of his craft, his friendly modesty is equalled only by his great skill.

A special award was made in 1973 to the widow of John Banks, an upland farmer who tragically died before any placings were made. At Ellershaw, Grewelthorpe, Ripon, Yorkshire, John Banks had re-seeded many of his difficult fields and yet, though very hard-working, was never too busy to help others. Many an awkward lambing was brought up Ellershaw's rough road to receive attention as skilled as any available. Neighbouring farmers showed their regard in truly practical fashion, joining to harvest all the silage at his farm.

A Dorset shepherd in sole charge of 2,000 ewes won the 1974 Award. Richard Lund, from a non-farming family, began working on farms while still at school. He spent two years as a farm student and one at Shropshire Farm Institute, having also studied farming at Hereford Technical College. He went from the hills to a lowland farm, where he progressively raised ewe numbers by a hundred a year.

During lambing Richard, his wife Sally and their two small sons move into a caravan parked at the lambing site. This area is designed to take an easy outward flow of ewes as soon as they have lambed.

As a schoolboy, runner-up Joe Tee helped his father dig buried sheep out of snow drifts on a Welsh mountain. The third generation of his family to be shepherd on the 900-acre home farm of the Rhiwlas Estate, Bala, Gwynedd, he has sole charge of 1,800 Welsh Mountain ewes plus followers.

According to his employer, Lt Col K. J. Price, prizes in the summer show ring totalling 150 rosettes and cards, plus top auction prices, speak volumes for the stockmanship and dedicated shepherding of Joe Tee.

When he was eighteen, the 1975 award winner, Bill Graham lost the use of one arm through a motor cycle accident, yet he can manage every job among sheep. 'We never hear Bill complain', said his employer Mr Lambert Carmichael, Scremerston Hill, Berwick-on-Tweed. 'Provided his sheep are all right, and he can get to sheep dog trials occasionally, he's happy. The big Suffolk-cross ewes that make up most of the flock of 1,000 are no problem to him.'

When Bill Graham had the accident which put his farming future in the balance, he approached Mr Carmichael to see if there was any job he could do, despite his recent disability. 'You can count sheep with one arm, Bill', was the reply, and his employer soon realised that here was a natural shepherd, with the makings of a first-class stockman.

A keen stockman will change jobs rather than give up the breed he knows best. When a well known Border Leicester flock was dispersed in 1959, 1975 joint runner-up Hamish Hall moved to Fishwick and the Stobo family, five miles inland from Berwick-on-Tweed. He has stayed there ever since. This was a prime factor in deciding his present employer James Stobo that Hamish Hall was a worthy Shepherd of the Year candidate. 'Both previous award winners were young, and I felt that recognition should be offered to older men who shepherded during the dark, dismal period of the 1960s, and have done a sterling job in the industry through thick and thin', he said.

The other runner-up was twenty-seven year old David James, a Welshman described by judges' chairman Professor Gordon Dickson as 'a young man with great keenness, energy and initiative'. Although the ewe flock at Sixpenny Handley, Salisbury, Wiltshire, runs into four figures, David turned in an outstanding lambing percentage, beating that of judge Harry Ridley on similar mixed arable country.

Shepherd of the Year, 1976 was, of course, Ray Dent whose example has been used throughout this book.

Another master shepherd, who became runner-up in the 1976 Shepherd of the Year Award, is James Purves, of Sandyknowe, Kelso, Roxburghshire. An inbye shepherd, he combines the highest traditional skills of the Border shepherd with a refreshingly progressive approach to his flock management, reported the panel of judges. James' success disproves the old adage that 'a sheep's worst enemy is another sheep'.

BIBLIOGRAPHY

Ark, The (monthly); Rare Breeds Survival Trust, Winkleigh, Devon

Bowen, Godfrey. *Wool Away!* (Whitcombe & Tombs, 1955)

British Sheep (National Sheep Association, 1976)

Cooper and Thomas. *Profitable Sheep Farming* (Farming Press, 1965)

Culley, George. *Observations on Livestock* (Robinson, 1794)

Firbank, T. *I Bought a Mountain* (Harrap, 1940)

Fraser, Allan. *Sheep Farming* (Crosby Lockwood, 1937)

Fraser and Stamp. *Sheep Husbandry and Diseases* (Crosby Lockwood, 1968)

Gossett, A. L. J. *Shepherds of Britain* (Constable, 1911) .

Grant, David and Hart, Edward. *Shepherds' Crooks and Walking Sticks* (Dalesman, 1972)

Hudson, W. H. *A Shepherd's Life* (Everyman, 1910)

Longton, Tim and Edward Hart. *The Sheep Dog; Its Work and Training* (David & Charles, 1976)

McCulloch, Herries. *Sheep Dogs and Their Masters* (Moray, 1938)

Moorhouse, Sydney. *The British Sheepdog* (Witherby, 1938)

Morrison, A. M. *Red Dragon Farm* (Faber, 1964)

Nixon, David, B. *Walk Soft in the Fold* (Chatto & Windus, 1977)

Ollivant, Alfred. *Owd Bob* (Heinemann, 1898)

BIBLIOGRAPHY

Perry, Richard. *I Went A-shepherding* (Lindsay Drummond, 1944)
Rainsford-Hannay, Col F. *Dry Stone Walling* (Faber, 1957)
Raistrick, Arthur. *The Pennine Walls* (Dalesman, 1946)
Rollinson, William. *Lakeland Walls* (Dalesman, 1972)
Robertson, R. B. *Of Sheep and Men* (Gollancz, 1960)
Ryder and Stephenson. *Wool Growth* (Academic Press, 1968)
Thomas. *Sheep Farming Today* (Faber, 1966)
Trow-Smith, Robert. *A History of British Livestock Husbandry*, 2 vols
  (Routledge & Kegan Paul, 1957)

ACKNOWLEDGEMENTS

The author wishes to express his thanks to: the Dent family—Ray, Lena and Ian—who helped crystallise his ideas during a protracted winter and backward spring; their neighbour, George Wall, who took the cover photo; Chatto & Windus; Faber & Faber Ltd; Heinemann Group of Publishers Ltd; Donald Dougal; Matt Mundell; Marcus Oliver and *Livestock Farming*; Bryn Roberts and *FMC News*; Peter Weston; Misses D. M. and E. M. Alderson, whose drawings are the breath of the countryside; Pamela Walker who typed the manuscript and Audrey Wickham who compiled the index.

INDEX

Page numbers in italic type refer to illustrations

125

INDEX

Ear tag, 60, 78
Enclosed grazing, *see* In-bye
Exmoor Horn, 28, 29
'Eye' in dogs, 39, 40

Fencing, 31
Fertility, 83
Fleece marks, 55, 57, 60, 64, *27*
Fleeces, 29, 55, 74
Flock mark, 72
Fly, *see* Blow fly
Footrot, 58, 103
Gathering, 38, 45, 55
George Hedley Award, 116–7
Glossary, 109–11
Greyface, 19, 35

Harris tweed, 20
Hay, feeding, 14, 15, 35, 56
Hay making, 16, 73, 88, 92
Heather, 85–6, *87*
Heft, 32, 84
Herdwick, 22, 30, 31, 54, 62, 81, 99, *24, 50*
Hirsel, 34
Hogg, 14, 16, 33, 64
Hormone action, 17
Horn, 95, 96, 97
Horned sheep, 19, 20, 26, 28, 53
Horn marks, 57, 62, 64, 74
Huntaway, 22

In-bye, 31, 35, 75
International Sheep Dog Society, 39, 114

Keeper, 44, 85, 87, 88

Lake District, Lakeland, 20, 22, 30, 31, 50, 88, 99, *54*
Lambing, 16, 31, 33, 35, 51, 52, 55, 67–73, 75, 82, 97–8, 104, 105, 106
Lamb sales, 16, 99, *100*
Lonk, 26

Masham, 24

Mating, 17
Merle dog, 40
Mothering-up, 74
Mountain rescue, 31
Mule, 21

National Trust, 30
New Zealand dogs, 22, 29
North Country Cheviot, 28, 31, *75*

'Pack' sheep, 108
Pay, 103
Peat, 87, 92, 93
Pedigree, 80
Pen, 46, 48, 55, 72, 73, 80
Penistone, 26
Points (trials), 45–6
Polled sheep, 22, 26, 28, 29
Potency, 83

Raddle, *see* Rud
Rainfall, 34, 37, 104
Ram, 77–84, 99, 100
Ram hiring, 22, 99
Ram sales, 21, 77–84, 94, 98
Register of Dykers, 83
Rig-welted, 28
Road mending, 16, 90, 91
Rough Fell, 24, 26
Rough grazing, 30, 31
Rud, 83

Scab, 56, 58
Scandinavian sheep, 22
Scottish Blackface, 19, 26, 29, 36, 49, 53, 81, 82, 107
Scottish Halfbred, 28, 36, 75
Scrapie, 83
Shanks, 95, 97
Shearing, 16, 45, 55, 73–6, 78, 83, 92, 104, 108, *54*
Shears, 53, 55, 73
Sheep dogs, *see* Dogs
Sheep dog commands, 42

INDEX